Praise for *Earth Grief: The Jou[rney]*

"This book is a plainspoken and shambolic masterpiece, moist with tears. It's the honest fruit of a life lived close to the rain, to the whispering of leaves, to the raggedness of the heart. Created by a man who has stopped each day to catch sight of, to chew on, and slowly digest his own shadow, it works a dark and joyous alchemy upon the soul of the reader."

– DAVID ABRAM, author of *The Spell of the Sensuous* and *Becoming Animal*.

"I don't know how Stephen has managed to awaken the feeling-wisdom in me with just 'mere' words, but boy – it works – and with such poise, grace, humility and insight. I will revisit this treasure of a book again and again to imbibe its medicine and remember how to feel most inwardly and how to respond most appropriately in my own life to the catastrophic realities of our times."

– STEPHAN HARDING PhD, Senior Lecturer and Deep Ecologist, Schumacher College. Author of *Animate Earth* and *Gaia Alchemy*.

"*Earth Grief* is a guide for navigating the turbulent and changing climate of mind that we have inherited today. When you are in unfamiliar lands filled with perils, it is helpful to find someone who has a feel for the terrain and has come to call that place home. Stephen Harrod Buhner steps into this elder role by telling stories that call us to struggle with our grief for what has died and is dying not only around us, but within us too. As he writes: 'while it is true that we need to change human behavior in the outer world it is even more true that we need to change human behavior in the interior world.' By bringing us face-to-face with such important truths, *Earth Grief* helps us reconnect our tears and heart-break to the sustenance we need for a different way of living."

– TIMOTHY B. LEDUC, Assoc Professor of Social Work, Wilfrid Laurier University, author of *A Canadian Climate of Mind* and *Climate, Culture, Change*

"This is a book for all those who feel the grief, pain and suffering of the Earth and our fellow-creatures – including many humans – and are tired of denying it, of trying to shut it out, and of all the false promises that there is a solution: a technique, a method, a system that will make it go away or be okay. It is a book about the central truth of our time: the collapse, already well underway, of the ecosystems that comprise life and support all human societies. Yet unlike so much other writing on the subject, it engages that truth with the deep feeling and emotional honesty that it entails and deserves. Buhner even offers hope (not mere optimism) that on the other side of the wrenching work of going down into the darkness of grief, both inner and outer, there might be a way to reinhabit our only home with integrity, humility and compassion. I know of no more important work."

– PATRICK CURRY, author of *Ecological Ethics* and editor-in-chief of *The Ecological Citizen*

What others are saying about Stephen Harrod Buhner's work

"One of America's preeminent herbalists, Stephen Buhner articulates the sacred underpinnings of the herbal world and deep ecology as only a real 'green man' can."

– DAVID HOFFMANN, Fellow of the National Institute of Medical Herbalists and author of *The Holistic Herbal*.

"Stephen Buhner's writings are a powerful call for people of all colors and nations to work together to restore recognition for and experience of the sacredness of Earth."

– BROOKE MEDICINE EAGLE, Native American teacher and author of *Buffalo Woman Comes Singing*.

"If Lao-Tzu and Emerson could have a dialogue . . . they would welcome the company of this remarkable book."

– WILLIAM HOWARTH, Professor of English, author of *Walking with Thoreau*

Also by Stephen Harrod Buhner

Nonfiction

Becoming Vegetalista, limited, hardcover edition of the first one-fourth of the book

Plant Intelligence and the Imaginal Realm: Beyond the Doors of Perception and Into the Dreaming of Earth

The Secret Teachings of Plants: The Intelligence of the Heart in the Direct Perception of Nature

The Lost Language of Plants: The Ecological Importance of Plant Medicines for Life on Earth

Sacred Plant Medicine: The Wisdom in Native American Herbalism

Sacred and Herbal Healing Beers: The Secrets of Ancient Fermentation

One Spirit Many Peoples

Language and Poetry

Ensouling Language: The Art of Nonfiction and the Writer's Life

The Taste of Wild Water: Poems and Stories Found While Walking in Woods

Ecological Medicine

Healing Lyme, second edition: Natural Healing of Lyme Borreliosis and the Coinfections Chlamydia, and Rickettsiosis

Healing Lyme Coinfections: Complementary and Holistic Treatments for Bartonella and Mycoplasma

Natural Treatments for Lyme Coinfections: Anaplasma, Babesia, and Ehrlichia

Herbal Antibiotics, second edition: Natural Alternatives for Treating Drug-resistant Bacteria

Herbal Antivirals, second edition: Natural Remedies for Emerging and Resistant Viral Infections

Pine Pollen: Ancient Medicine for a New Millenium

The Transformational Power of Fasting (previous incarnation: *The Fasting Path*)

Natural Remedies for Low Testosterone (previous incarnation: *The Natural Testosterone Plan*)

Herbs for Hepatitis C and the Liver

Vital Man

Earth Grief

THE JOURNEY INTO AND THROUGH ECOLOGICAL LOSS

Stephen Harrod Buhner

Raven Press
Boulder, CO

For information on this book,
please contact the author:
Stephen Buhner
8 Pioneer Road
Silver City, NM 88061

Copyright © 2022 by Stephen Harrod Buhner
All rights reserved. No part of this book may be reproduced or utilized in any form or by any means, electronic or mechanical, including photocopying, recording, or by any information storage and retrieval system, without permission in writing from the publisher.

Publisher's Cataloging-in-Publication data

Names: Buhner, Stephen Harrod, author.
Title: Earth grief: the journey into and through ecological loss / Stephen Harrod Buhner.
Description: Includes bibliographical references. | Boulder, CO: Raven Press, 2022.
Identifiers: ISBN: 978-0-9708696-7-8 (paperback)
Subjects: LCSH Climatic changes. | Environmental degradation--Psychological aspects. | Ecological disturbances--Psychological aspects. | Global environmental change--Psychological aspects. | Grief. | Anxiety. | Loss (Psychology). | Human beings--Effect of climate on. | BISAC NATURE / General | NATURE / Weather | PSYCHOLOGY / Grief & Loss | SELF-HELP / Death, Grief, Bereavement
Classification: LCC GE140 .B84 2022 | DDC 363.7--dc23

ISBN 978-0-9708696-7-8

Printed and bound in the United States of America at TK
15 14 13 12 11 10 9 8 7 6 5 4 3 2 1

Text design and layout by Sterling Hill Productions

Thanks are extended to:

Robert and Ruth Bly for permission to reprint Robert's translations of one poem each by Machado and Jimenez as well as his own work in this book.

William Stafford, "A Ritual to Read to Each Other" from *The Way It is: New and Selected Poems*. Copyright © 1960, 1998 by William Stafford and the Estate of William Stafford. Reprinted by permission of The Permissions Company, Inc. On behalf of Graywolf Press, Minneapolis, MN, www.graywolfpress.org and the William Stafford Estate.

"A Song on the End of the World" from *The Collected Poems* by Czeslaw Milosz. Copyright © 1988 by Czeslaw Milosz Royalties, Inc. Used by permission of HarperCollins Publishers.

Dedication

For all those who feel the grief of Earth

*We must ask ourselves:
What has become of the garden
that was entrusted to us?*

Contents

The Beginning: In Which the Author Positions Himself	1
Chapter One: The Journey before Us	11
Interlude One: The Teaching of Barns and Birds	17
Chapter Two: Earth Grief	20
Interlude Two: "Once You Are Real You Can't Become Unreal Again"	46
Chapter Three: Shame and Guilt	49
Interlude Three: "People Possess Four Things That Are No Good at Sea"	83
Chapter Four-The-First-Part: Inevitability and Descent	91
Narratio Interruptus: The Diagnosis	116
Chapter Four-The-Second-Part: Inevitability and Descent	163
Interlude Four: Fragments from a Stained Glass Window	176
Chapter Five: The Journey through Grief and Loss	191
Interlude Five: "It's Hard to Live without Love, You Know"	230
Chapter Six: The Life We Live with Afterward	241
Epilogue: "It's Hollow Inside, Isn't It, Just Like Me?"	251
Bibliography	270

THE BEGINNING

In Which the Author Positions Himself

I hold to the presupposition that our loss of the sense of aesthetic unity was, quite simply, an epistemological mistake. I believe that that mistake may be more serious than all the minor insanities that characterized those older epistemologies which agreed upon fundamental unity.

GREGORY BATESON

Educating the mind without educating the heart is no education at all.

ARISTOTLE

There must be those among whom we can sit down and weep and still be counted as warriors.

ADRIENNE RICH

I have spent more than a half century as an activist for Earth and for sustainable human habitation of our home. My education has been as broad as I could make it for I had a feeling, even at the beginning, that only a holistic, multidisciplinary training could possibly give me the foundation necessary to do the work that I felt called to do. I had no idea, as none of us do, of course, of what I was getting myself into. I suppose if any of us did, none of us would ever do anything.

My education, and the real-world training it demanded, has involved decades of study and experiential exploration in fields as diverse as tree ecology and communication, contemplative spirituality, plant movement patterns and function in ecological space, human psychology, Gaian emergence and behavior, death and dying, indigenous land relationships, a wide range of psychotherapeutic techniques (including years of my own psychotherapy and work as a psychotherapist), the function of plant chemicals in ecological homeodynamis, animism, decades of work and training in rehabilitating broken homes, the nature of language and meaning, the art of storytelling, symbiotic relationships in ecological systems, the long term history of widely divergent human cultures, the origin and history of christianity, nonlinearity in complex systems, various forms of human and ecosystem healing, rationalist science and its impacts over time on culture and ecological structures, cytokine dynamics in disease, plant/microbial ecological interactions/genetic and chemical alterations and responses, decades of work (and training) as a teacher and public speaker, the nature and intelligence of human organ systems, the fall of Rome, the ecological function of plant medicines, the impact of pharmaceuticals on ecosystems and their functioning, fine woodworking, the dynamics of civilizational collapse, symbiogenesis, decades of study and work as a writer and publisher, close examination of western medical and scientific systems, microbiomes and their ecological function, tonal dynamics and patterns in music, transcultural epistemology, the ecological importance and function of "invasive" plants, the function of psychotropics in ecosystems, phytoremediation, mathematics, decades of work (and training) as a community herbalist, the drug war, the cultural impact of prohibition laws (of whatever sort), the economic and cultural impacts and outcomes from professional licensure, plant and bacterial ecology and communication in ecosystems, neural network dynamics, neurogenesis, sound patterning in language, the harvesting and preparation of wild plant medicines, deep and long term experiential work with a great many different states of mind/consciousness/being, and a great many more things as well. Over the decades these have integrated themselves into a whole that has become far more than the sum of the parts – which they had to do if I was going to be able to do the work that I do.

Yet, something far more important than all of this study and training is foundational to my work. Long before I sought out those separate disciplines, spent half a century learning what they had to teach me, something powerful enough to send me on such a journey was already at work within me. And that is a *feeling* thing, not a thinking thing. It is love and caring. Not only for the various disciplines, healing modalities, and artisanal work I have pursued but for our home, this Earth, for all the kindred species that surround us. It is a passionate love that comes from the deepest core of the self. It is the most important thing of all. And it has been felt by, and motivated, hundreds of millions of human beings since human beings have been. Yet it is the one thing that is nearly always left out of scientific and policy discussions of what to do to respond to climate change. And it's rarely acknowledged as a crucial aspect of the individual disciplines I have spent so many years learning.

One of the deep-seated, and very dangerous, problems we face is the tendency for rationalists (of whatever sort) to leave out love – and the empathy and caring that comes with it. And a science without love and empathy, a technology without love and empathy, a way of thinking and being without love and empathy . . . well, we see the results of that around us every day of our lives.

Tellingly, authors of many of the better research papers and academic books I have read feel a need to apologize for having feelings about what they are studying and for what they see – and feel – happening around them. They have internalized rationalist shaming about their feelings of love and caring for life forms external to themselves, about the grief they feel when the trees they study, and love, are cut down. When they touch upon those feelings in their writing or in their public speaking they experience an almost immediate, internal, ingrained denigration of their emotional responses – and themselves. And so, nearly all of them believe they must pretend to have no feelings at all, as if somehow this will make their work more legitimate. Phyllis Windle, in her remarkable article "The ecology of grief," describes it like this . . .

> *I suspect that ecologists, like other scientists, are prone to inhibiting the pain of grief. We are solidly attached to the life of the*

> *mind and, of the several steps experts consider essential to recovery, only the first is intellectual.*
>
> *I speak from experience. I am tempted to dismiss my feelings for dogwoods as irrational, inappropriate, anthropomorphic. My arguments go like this: another tree will take the dogwoods' place; death is part of productivity, too; evolution removes as well as adds species. . . . [But] premature reassurance and pressure to accept a loss just short-circuit the grieving and recovery process. . . .*
>
> *We have almost no social support for expressing this grief. . . . Honest conversations about grief that come quite naturally at a bedside are far more difficult at a lab bench or conference table. Thus, it is harder for me to speak freely about my grief for dogwoods with ecological colleagues than with fellow chaplains.*

She's identified something crucial to the problems of our time. It is, in fact, integral to the emergence of those problems. More plainly, it's one of the main causes of the problems we face. It's the insistence among scientists (and the schools and teachers who train them) that our natural human feelings of caring and grief have no place in their world nor in any legitimate understanding of the natural world itself – that they do not belong in the practice of science.

> *The insistence that caring for the outward world and the kindred life forms that surround us is "irrational, inappropriate, anthropomorphic" is, in fact, pathological. To be specific: psychopathic personality disorder is defined by an inability to feel love, a lack of empathy and remorse, shallow affect, glibness, callousness, and the manipulation of others (which for me includes life forms other than our own). These are all elements of the mindset inculcated in scientists (of whatever sort) who study the outward world. Thus it is considered wrong, misplaced, un-scientific to feel love and empathy for what is being studied, remorse for its loss, discomfort with its*

manipulation or callous treatment, for anyone to act as if life forms other than our own have inherent dignity and the right to be treated as one would treat a human being, or to give up "objective distance" and instead develop an intimacy with the outward world and its non-human inhabitants. These behaviors, it is believed, have no place in science. Those who do so despite this are told they are "projecting" human attributes onto an outward reality that does not possess them and their work is then denigrated by their peers.

But what is really true is that whenever a person is told these things, that their feelings of love and caring for Earth, the trees, for all life is wrong, irrational, inappropriate, or anthropomorphic, they are being urged to conform to a pathological, psychopathic mindset. And that mindset is, by its nature, anti-life and anti-human. Love and caring are essential to our humanness, to our successful habitation – and understanding – of Earth. And our grief? It is an ancient, evolution-developed signal that something essential is being lost. We ignore it at our peril.

The *imperative of the impersonal* has become the default orientation of science, of rationality itself. It hasn't always been this way (not that people have not always had problems or behaved stupidly). There are *other* kinds of science – and rationality – in the world. (And that's a strange thought, isn't it? That there are other kinds of rationality, other kinds of science.)

It took years before I understood that what I was training myself in was a comprehensive, holistic, and contemporary expression of what used to be called natural philosophy . . . what eventually became, in a very reductionistic and badly distorted form, the thing we now call science. Before its subversion, love and caring, wonder and joy, were part of the study of natural systems – and the lives of the people involved in it. (And certainly no one ever apologized for it or felt ashamed to talk about its importance in their lives and work.) If you read any of their work (and I have) it is clear, and often stated, that love and caring, wonder and joy, were felt to be essential aspects of the kind of science they did. Many of the early natural philosophers warned against the strain of science which believed that rationalist dissociation, an impersonal objectivity, and dissecting the world into parts was the proper way to understand nature. They knew that Earth was far more than insentient matter, that there was a great deal more going on here than a mechanicalistic materialism could ever explain.

Nevertheless, those who pursued dissociation from and dissection of the world believed, insisted actually, that there was no intelligence out there, no soul, nothing to personally relate to, just insentient matter. And somewhere along the way, this group of people, the ones I consider to be the most psychologically damaged of the natural philosophers, gained dominance in the field. Once they did, they set about making sure that everyone who came after them saw the world in that way as well. Then they set about making sure that the rest of humanity believed it, too. But of course they forgot the most important thing (or maybe they were just afraid of feelings), and that is that those things we do not care for we will not be care-full with and will, as a matter of habit, become care-less about. More obviously, even reductionists should have realized that the dissection of living systems kills what is dissected, or, as Aldo Leopold once put it, "The first rule of intelligent tinkering is to save all the parts." But then again, as James Hillman said, maybe the whole point for them *was* the killing, for: "It was only when science convinced us that nature was dead that it could begin its autopsy in earnest."

It's a hard thing for those of us who grew up believing in and loving science to come to the difficult emotional conclusion that the way nearly all scientists go about their work is a primary source of damage to our world.

It is the discoveries of scientists, and their objective dissections of the natural world, that have given technologists and industrialists the information they needed to disassemble the ecosystems of our planetary home. We live amidst the ruins that their work and way of thinking have made of the natural world. It's unlikely that the same kind of thinking and acting that got us into this mess will get us out. I have never seen an instance where the more you drink the soberer you become. Something else, a different approach, is needed. And that is what all of my work, including this book, is about.

As the years went by I came to understand that there was an additional, and crucial, element of the work I was engaged in. And that is, as a tribal elder I respect once put it, it has to "grow corn." Specifically, all that high-fallutin' language and study don't mean diddly if it can't be brought into the world effectively – here – on the ground – where *all* the living beings of this planet live. (Too many academics, as Bill Mollison once put it, never have dirt under their fingernails.)

Growing corn means actually doing something in the real world; it doesn't just consist of a lot of pretty talk. (And it involves a great deal more than the dissection and manipulation of nature.) It entails engaging in long-term work in the real world, in real ecosystems, constantly learning to refine the shape and expression of the corn that is being grown. (And by this I am definitely *not* talking about the creation of genetically modified organisms; that is the antithesis of the point.) It means actually finding out if the pretty theories have anything to do with the real world and, when they are brought into practice, actually looking at how they affect people's lives, the lives of our nonhuman companions, and the health of Earth's ecosystems – it means that we have a responsibility for the ripples that our work sends outward into the world. We can't, for instance, send physicians into a country to reduce infant mortality without also taking into account the impacts that an increase in population will have there once we are gone. That's no different than leaving the ecological ruins of an open pit mine for later generations to deal with. In other words, the corn we grow has to be grown in very specific ways.

Growing corn, the way that tribal elder meant, is the same thing that Aldo Leopold was talking about when he struggled with the problems of developing a coherent land ethic that can guide us in distinguishing ecological right from wrong. Leopold said this: "A thing is right when it tends to preserve the integrity, stability, and beauty of the biotic community. It is wrong when it tends otherwise." And I would add this: *it is right when it acts to bring into balance that which has been destabilized by wrong action.* Approaching our work this way creates moral parameters around the corn we grow, limits our behavior in specific ways, forces us upon a particular kind of road. (And how many scientists, industrialists, or physicians have ever asked themselves whether, by this definition, the actions they take, the work they do, is ethical? I have met hundreds if not thousands of them; the answer is very few. Too many of them say, "I was just doing the science, I am not responsible for what people do with it afterwards.")

Growing corn is always a very simple thing – it rarely involves four-syllable words or a lot of fancy theories. It's something done in relationship with the livingness of the world. It's a co-creative process – not a monologue but a conversation. And now a basic problem in the dominant approach emerges – the only way to have the kind of conversation I am talking about is to acknowledge that there are intelligences here other than the human – that there is someone on the other side of what is being said, someone who can both hear and respond. Or as Barbara McClintock once put it, "I went no place that the corn did not first tell me to go."

Only if we listen can we hear what Earth and our hearts are telling us. Only if we listen can we learn how to grow the corn that we, as individuals, are uniquely meant to grow, corn that is ethical, that "preserve[s] the integrity, stability, and beauty of the biotic community," whose growing "acts to bring into balance again that which has been destabilized by wrong action."

And finally, there is this . . . "corn" is a metaphor for every action we take, every solution we suggest or implement. Earth and many of our best teachers tend to speak, to teach, in metaphors. There is a reason for this (and I will talk about it more as this story goes on). Literalism is and has always been one of the most dangerous paths for human beings to take. (If you are a literalist, I don't think you will like this book – or anything I say in it.)

The problems we face are incredibly complex and, at this point in time, massively large in scope. Attempting to change even a single aspect of what we face is incredibly difficult. As many long-term, ecological activists have learned, we (and the work we do) are very similar to the rudder on the *Titanic*. Rudders are such tiny things compared to the size of the *Titanic* – their effects can only occur over long timelines. Nor are they always successful at altering course in time.

It is inescapable then that during our long years of Earth work, as the early youthful exuberance and lack of sophistication with complex systems (including the human) mature through decades of exposure to those complexities, that the size of the problems we face becomes ever clearer to the eye. It slowly becomes clear just how little we can actually do to change what is happening – for the inertia of the systems in play is far greater than any of us understand when we begin our work. As Aldo Leopold also said, "One of the penalties of an ecological education is that one lives in a world of wounds." There comes a time, for all of us, when we finally, experientially, in our feelings and our intellect, grasp the size and pervasiveness of the wounds that surround us, when we finally grasp the immensity of the work that lies before the human species. That realization is very hard to bear.

A fundamental truth is that as our education continues, the more of those wounds we will see and feel. After a while it becomes very difficult to emotionally endure them. The years go by and nothing much seems to change, the patterns of Earth destruction continue and grow ever more extreme as human population and corporate power increase – sooner or later most of us hit the wall. We suddenly find ourselves surrounded by the darkness that is always waiting there, just beyond the edge of the light. The emotional cost of Earth work is high, burnout a constant possibility, and periods of despair and depression inevitable.

I have burned out several times, experienced deep and dark depressions, struggled with a terribly wounded heart, felt constant pain from the ecological damage I experience around me every day of my life. It has taken decades for me to find the solutions to these things – to find the climate of mind and heart that Earth and the work have been demanding of me.

Not so surprisingly, the solutions I found were not where the courtiers of power, the licensed or degreed, insisted they were. I found them someplace else, a place only those who listen to the call of Earth tend to travel – out there, far out there, in the wildness of the world.

The answers I found are what this book is about.

CHAPTER ONE

The Journey Before Us

For one species to mourn the death of another is a new thing under the sun.

ALDO LEOPOLD

I found that to tell the truth is the hardest thing on Earth, harder than fighting in a war, harder than taking part in a revolution. If you try it you will find that at times sweat will break upon you. You will find that even if you succeed in discounting the attitudes of others to you and your life, you will wrestle with yourself most of all, fight with yourself, for there will surge up in you a strong desire to alter facts, to dress up your feelings. You'll find that there are many things you don't want to admit about yourself and others.

RICHARD WRIGHT

Grief will be our companion on this journey – it is not something we can deal with and move on.

LESLEY HEAD

If you spend any time reading about climategrief/ecologicalgrief/solastalgia, which are some of the words/terms/descriptions that people are creating for what so many of us are feeling, you will find a great many superficial suggestions of what to do about that feeling: talk to a therapist, take antidepressants, keep a journal, take anti-anxiety medications, join support groups, take anti-psychotics, change your work, sign yourself into a treatment center, write your congressman, take action (so you will feel as if you have personal autonomy and control again), meet with others who feel as you do (so you won't feel alone in these feelings), find something to feel optimistic about, and so on. All these suggestions have one thing in common: they are intended as a means to diminish, end, or transcend the feelings of grief, pain, depression, and hopelessness that increasing numbers of people are struggling with. They are interventions of one sort or another that, it is hoped, will stop or reduce the pain of the terrible grief so many of us are feeling (and perhaps, as well, to make you a "productive" member of society again). Despite the fact that some of these things may be useful during various stages of the journey (and I am not saying they are not), I am suggesting a different path.

Rather than suppressing (or strategically avoiding) the feelings of grief and pain in order to feel somewhat normal again (making it possible to go on as we have been going for far too long), the solution just might lie someplace else. Healing, and the wholeness that comes from it, is far different than what is found by suppressing pain – for pain is only a response to an underlying problem. It's a signal. Its function is to direct our attention to where the problem originates. Suppressing the pain without accurately identifying its source and then crafting an intervention to address it just allows the damage to continue, to, over time, become worse, to, perhaps, become terminal. People are pretty good at dealing with splinters from a piece of shattered wood, not so good at dealing with broken hearts from a shattered world.

In fact, the tendency to suppress or avoid rather than deal with the underlying condition that our pain and grief are a signal about – that is, the orientation of mind out of which the tendency to suppress or avoid comes – is itself part of the problem. It is embedded within what you might call a particular *climate of mind*. And that climate of mind is entangled far more than most people realize with the climate problems we face. What is true is that it is the climate of mind *in here*, inside us, that has given rise to the climate problems *out there*. Maybe it's time to go a bit deeper, to do something else. Perhaps to do the opposite of what we are most often wanting or being urged to do.

The extensive destabilization of Earth's ecosystems that now threatens human habitation of our planet did not happen in a vacuum. It has come from particular states of mind, mental software, orientations of self and culture that are and have always been inaccurate to the complexity of the nonlinear, self-organizing ecosphere of Earth. Over the last two millennia, unrestrained expression of those states of mind through our now eight-billion-humans behavior *is* the problem. And since those states of mind and being are connected at the core to primal, unthinking survival drives in each of us, stepping outside them, for nations or individuals, is very difficult indeed.

So . . . you won't find those particular suggestions in this book – the grief, pain, depression, and hopelessness we feel are not the problem (despite the soul-shattering impact of those feelings). They are *symptoms* of the problem. And it is the problem itself that we must face if we wish to address the crisis of our times . . . a crisis that exists out there in the wildness of the world as well in here in the secret chambers of our own hearts.

You should know, right here at the beginning, that the journey I suggest each of us take to effectively face what is in front of us is not an easy one – nor will most "experts" agree with it. In fact, it will cost you everything you think you are. (If it were easy everyone would already have done it.) But then again, what the planet is facing is not all that easy either and I am pretty sure that Earth is not going to let us off the hook if we simply spend a couple of months talking to a licensed therapist so that we feel better.

What I am suggesting in this book is something quite different. And I will tell you here that yes, if you truly want it, you will find the resolution/solution you seek – it's just that it will look very different than anything you have imagined, for it lies outside current ways of thinking and being and living and that's really the point. We will not solve the problems we face using the same kinds of thinking/being/living that got us into this mess to begin with.

To begin with, you need to understand that the feelings we have (and yes, I have them, too) are a response to a communication from the heart of Earth. We humans, like all living things, have been expressed out of this ever-moving, ever-changing scenario that we call Earth. We are irremovably, inextricably embedded within that scenario. And it is important to understand that it *is* a scenario; it's not a *place*. (The erroneous belief that

Earth is a place is a significant aspect of the problems we face – it's integral to the mind/body dualistic split and as such is foundational to the destruction of the natural world.) As Buckminster Fuller once put it, "It is possible to get out of a place, it is not possible to get out of a scenario." (Or as he also said, "Earth is a verb, not a noun.") The real world is far different than what we have been taught.

We are Earth itself, expressed into a particular form out of a living field of meaning which, physically, is an ever-changing, ever-adapting ecological matrix. And we are inextricably embedded within that living scenario, an irremovable part of its fabric. We are not, as we have too often been told, isolated consciousnesses inhabiting a ball of resources hurtling around the sun. Thus when Earth moves/communicates/changes, everything that has been expressed out of Earth feels it. For despite the centuries-long pronouncements of monotheists and rationalists to the contrary, what is actually true is that everything that is, even Earth itself, possesses *language*, communicability, and intelligence. It's just different than how *we* think of those things.

> *And a crucial part of the problem we face is that for the first time in the history of human habitation of Earth, we, as a species, have decided to ignore this truth about the livingness and intelligence of Earth – something that all of us once knew as children and that all ancient and indigenous cultures knew/know as well.*

Nevertheless, the *language* of Earth and every one of its inextricably interwoven expressions (which we call plants and animals and insects, bacteria and viruses and fungi, i.e., life) is far more complex and interesting than the simple meaning-structures our species has created and which we call words. (Just because something does not know or use the word *ratiocination* does not mean that it cannot think or speak.) And the complex emotional state that I call *Earth Grief* – and that others call climategrief/ecologicalgrief/solastalgia – is our *feeling* response to a communication from the heart of Earth, urging us to take a path, and engage in a work, that is quite different than the one our species has taken/engaged in the past two thousand years. We are being called, individually and collectively, to reinhabit our interbeing with the world.

The feeling impact of this potent Earth message captures our attention in a way that "Hey! You!" will never do. For it bypasses the rationalist mind and touches something much deeper in the self, the place where the self most essentially *I* or *you* or *we* lives. It is the self we rarely show to others, the self most of us do not listen to when it comes time to decide the course of our lives, the self that is far more powerful when it takes action than the "I" that lives up here in the brain and that most of us have been taught is the only "I" there is. And as that potent Earth message touches our depths, its meaning ingested by the deep self, we feel grief, a grief more powerful than words can express. Because of this the "I" that lives up here in daylight cannot, as it usually does, avoid what the deep self is telling it, cannot escape what is now being felt. And so it must, in one way or another, begin to respond to what has so strongly captured its attention. (Though, of course, as is always true for each and every one of us, the first, most immediate, response is to find a way to make the uncomfortable feelings go away.)

This powerful touch of Earth grief is asking us to turn our face toward what is happening (rather than away) and in so doing take a particular kind of journey. And that journey has never been more important than it is now when we find ourselves balanced precariously between the old ways that are ending and the new ways that are struggling to come into being. Regrettably, the only way to make that journey is to travel into and through the grief that is being felt. The answer lies not in making the painful feelings of grief go away but rather in turning toward them and traveling more deeply inside them. It means finding and then facing the *cause* of the pain – and that cause is much deeper and more complex than "they cut down the trees."

It is a journey that all of us inevitably take alone – and that is one of the things that is so difficult about it. But I have taken that journey just as others before me have done. And just so you know, it is survivable – though when you are in the middle of it, it seems it will never end, as if nothing of use is happening, that no healing is occurring. However, what is really true is that it will end, much is happening that you cannot see, and the healing you seek really is happening.

Astonishingly enough, the decision to turn the face to the source of the pain and grief, to fully embrace it, stimulates, over time, the emergence of the form of Earth work that is uniquely yours to do: work that comes from your essential genius, the work you were born to do, the work that Earth needs you and only you to do. And this is crucial: for it is each of us in our millions, doing the unique work that is ours alone to do, in the particular landscape we inhabit, that is the solution to the crisis of our times.

On the other side is an integration that is essential to our times. For you will then hold in your body, your self, your presence, and the words you speak the emotional and existential resolution to the crisis we now face (which in and of itself will make a difference to those around you). Nevertheless, getting to the other side of all the pain and grief is difficult; it takes a great deal of internal work and much self-examination. Again . . . it isn't easy.

> *You should know that this journey . . .*
> *it's like peeling an onion.*
> *You remove the layers one at a time,*
> *and sometimes you weep.*

But it's important to remember: many people have taken this journey before you. They traveled in this kind of pain and darkness, too, and they found the way through. Remembering that, *as an article of faith,* is one of the few things that will help you through when times are hardest, when you think that you just can't go on.

It has saved me more times than I can say. For I *believed* the words that those who went before left for me and for all those who followed after. I had faith in them and I trusted what they told me. In the darkness their presence and the words they'd left behind came to me, uplifted me, and helped me endure. And so I continued on, no matter how hard it was on that particular day or week or month or year. As Walt Whitman once put it, "Be not discouraged, keep on, there are divine things well envelop'd, I swear to you there are divine things more beautiful than words can tell."

INTERLUDE ONE

The Teaching of Barns and Birds

The ship which keeps life afloat is constructed of surprisingly flimsy material. Not all shipwrecks are physical; there are things which can swallow up what is inside you as easily as the sea.

LORI FOX

You have an utterly sincere glance when you look around after you are shipwrecked. You weep a long time on an island when you realize what's happened to you.

ROBERT BLY

I wish I could name the strangeness of grief.

LECH BLAINE

Over time, I have learned that I'm *always* in the midst of the answers I seek. Yet I'm often so busy with my own struggles or the demands of daily life that I miss them for years and years and years. When I do finally notice what's been right in front of me all that time I'm always surprised at my lengthy inability to see, the depth and breadth of my ignorance – and how obvious the answers I've sought turn out to be.

There's a humility that comes from this – but it takes years to develop (my arrogance was strong and deep and unremitting). The point (I've found) is to not mind being ignorant while at the same time doing the work necessary to develop the skill of seeing what is right in front of me, all the while realizing that my degree of ignorance will never lessen (for even if I see and understand a million lessons, life has trillions more I will never notice). More simply, years of living sanded the sharpness off me until all of that pride was gone. Only then did it become easier to see what's been right in front of me all along. All of us are fools, it just takes a while to not mind.

That pride is part of the reason it took me so long to understand that the natural world most often speaks in metaphors. And this is something very hard to grasp for those of us who have been taught to be literalists. As the poet Dale Pendell so eloquently put it in one of his poems (of which this is a part):

> *Though the gods have the power of speech*
> *more often they choose a flower or plant;*
> *elder leaves pressed on a blotter,*
> *or spring buds emerging from a winter stem.*
>
> *These messages they send –*
> *so ordinary we often miss them:*
> *an easy laughter and lightness,*
> *or legs casually crossed and touching,*
>
> *The way a serpentine dike*
> *blends seamlessly into bedrock*
> *or the way two possible lovers move,*
> *starting and stopping, passing and pausing*
> *on an April trail.*
>
> *The subtlest oracles are always the most obvious –*
> *seeing what's in front of us the most difficult:*
> *a butterfly hatching from a ruptured dream,*
> *or a splintered tree rooting in the soil where it fell –*

It has taken decades of living for me to learn to understand these kinds of metaphorical communications. And yet, the more I do, the more easily I discover the places that we, as a species, are pinned – and the more easily

I find the answers I seek. One of the most important answers I needed came from the poet Robert Bly, who pointed out something I had noticed many times but not *seen* – for in those younger years I was still thinking too literally (something that I think comes from pride and a desire to control more than anything else – though in my youth this was very hard to understand).

I was born in 1952 and over the years of my childhood and young adulthood I spent a lot of time in old barns . . . in Kentucky, in Indiana, and, eventually, in the Colorado high country that became my home for so very long. And as was always true (before the days of metal barns), the barns I explored (each filled with the secrets and smells that only old barns know) were sheathed in wooden boards. Over time those boards (as all boards do) had grown old and weathered (much as I have done). They wrinkled and shrank and gaps appeared. And as always happens sooner or later (for that's the way birds are), birds crept through those gaps into the interior of the barns. Once inside they would panic (as birds often do) when they realized they could not easily find their way out again. They would fly up and up and up, battering themselves against the strips of light shining through the gaps in the boards of the barn. I would find them sometimes, days or weeks later, dead, often desiccated, on the floor of the barn. It was just one of the things that is, a part of life, like the moon and the rain.

Robert Bly also noticed these things (for he lived much of his life on farms in minnesota). But he saw something I didn't. Down along the floor of barns, rats and mice had gnawed at the boards, widening the cracks, so they could get into the barn and eat the grain that had fallen from the cattle and horses as they ate. And so it was he that found the teaching I had missed.

> *The way out*
> *is not upward*
> *and into the light*
> *but downward,*
> *into and through*
> *the darkness.*

CHAPTER TWO

Earth Grief

Grief is an acute state of awareness in which the fragility of the world reveals itself.

Suzanne Moore

It seems to me, that if we love, we grieve. That's the deal. That's the pact. Grief and love are forever intertwined. Grief is the terrible reminder of the depths of our love and, like love, grief is non-negotiable. There is a vastness of grief that overwhelms our minuscule selves. We are tiny, trembling clusters of atoms subsumed within grief's awesome presence. It occupies the core of our being and extends through our fingers to the limits of the universe. Within that whirling gyre all manner of madnesses exist; ghosts and spirits and dream visitations, and everything else that we, in our anguish, will into existence. These are precious gifts that are as valid and as real as we need them to be. They are the spirit guides that lead us out of the darkness.

Nick Cave

On Earth there is no escape, no exit, from global ecology.

Dorion Sagan

Grief, bereavement, heartbreak, despair, melancholy, depression, rage, fear, anxiety, terror, confusion, bewilderment, insecurity, hopelessness, bleakness, helplessness, resignation, fatalism, meaninglessness, emptiness, overwhelm, burnout, breakdown, pain, suffering, illness, death, apathy, lethargy, listlessness, frustration, aggression, hatred, self-condemnation, self-loathing, pessimism, cynicism, shame, guilt.

I am sure there are more, but these are the states of being and mind and heart that have made their home inside me at one time or another during the years of my activism. Every one of us who loves Earth encounters them sooner or later. They are each an element of our emotional response to the field, the signal being given off by the damaged ecosystems around us, by the destruction of what we love, the dying of that which has given our species its existence, which has birthed each and every one of us. We each respond to that signal in our own way – the complex of emotions we feel is unique to our nature . . . and to our character.

Very few people speak of this as openly and immediately as their grief needs them to . . . yet every so often someone realizes that, finally, they must before the pain of it consumes them. And so they unconceal themselves, reveal to the outward eye and ear that which their heart will no longer let them ignore. Ashlee Cunsolo is one who has done so, in her book *Mourning Nature:*

> In 2011, I was bereft, I was at a loss intellectually and emotionally. I felt adrift in waves of sadness, grief, loss, and pain, unanchored from my life, isolated from those around me, and unsure of how to process what I was experiencing. It was a sense of almost abyssal sorrow, without an idea of how to move forward, or how to see beyond the edges, the fringes of these feelings. It was a grief I did not expect or anticipate. Yet it was there waiting for me in the morning when I awoke, there at night as I drifted off, in my dreams and a constant companion throughout the day.

And so, too, has the pseudonymous Nightshade at her blog *The Purple Broom* . . .

> It is a gut wrenching feeling, seeing the world around me become so much more broken. This pain is overwhelming. It leaves me physically sick. I find I often burst into tears when the radio or

> *the news are playing in the background. I sometimes have to turn away from Facebook and the posts from my dear friends who post about the nuclear power plant and the dolphins being hunted. I cannot like, or react to these posts. I have to skim over many of them. I cannot see an animal in pain and not burst into tears. I cannot see anyone harming another being without becoming storm and water.*

And here is one of mine from among the thousands of grief events that live inside me, that will always live inside me . . .

During the first year of the trump presidency, when the government shut down after a budget impasse, the rangers in the Gila national forest were put on leave. Soon afterward the battered trucks came, carrying the young men in their worn clothes, their faces marked with privation and fear and the long years of their poverty. They stopped along the sides of the narrow mountain road that wanders up into the Gila, pulled out their chain saws, and began to cut.

The people in this region don't use pine to heat their houses (even though it works just fine); they prefer the oaks and the junipers – the oaks for their density, the junipers for their smell. So the young men stopped their trucks by the ancient alligator junipers that lined that mountain road and there they unlimbered their saws. The wood they took wasn't to heat their homes but to sell. A single tree, a century to a century and a half old, brings perhaps $250.

Unaware, I drove the road a few days later. I knew those trees, they had been my companions for a decade, friendly sentinels I'd said hello to whenever I drove deep into the Gila, to the place our tiny cabin rests, in the valley where Sapillo creek runs, where the ancient oaks and ponderosa pines grow, trees that were young when the apaches knew this land.

I remember the first one I came upon, its stump now a bare six inches above the ground. The trunk was a good two and a half feet in diameter, the discarded, scattered limbs evidence of the crime. Then I saw the second, then the third. Every mile or so for the next twenty miles there was another, then another, another, another, another.

A pain took root inside, wove itself into me; it will be with me until I die. That pain reached into the depths of me, tearing out the me that had been and shredding it onto that mountain road, mile after mile after mile.

And the pain? Two giant hands grabbing hold of my heart as if it were a cloth and twisting it, harder, then harder still until the pain is more terrible than my words can tell. It's the third-grade bully and his friends who caught me that day on the long walk home; the way they held me as he drove his fist deep into my stomach; the way they laughed as I lay on the ground, the way they swaggered away. It's the way I can't catch my breath, the loneliness of being nine and weak and lying on an empty street, my home far away, feeling I would never be safe or warm or loved ever again. It's the day I put my beloved dog to sleep, my friend of seventeen years; the way he looked up at me – those sad, pained eyes – as I held him, as the needle went in, as the plunger went down, the way he shuddered, and that tiny, almost infinitesimal moment when he took that last breath and was gone. It's the ache I live with every day, the empty hole he'll never fill again. The emptiness of never again seeing his happy face as I come in the door, never again seeing him pick up the ball and asking me to play, never again having him lay his head upon my arm as I lie there reading in bed, never again feeling the love come off him as he looks at me, never again feeling the love flow out of me and cross the space between us where it always found its place inside him, never again feeling the way our mutual love settled down inside us, like a dog circling its bed, or the way the deeps of us sighed and felt at home and wanted and loved when it did.

The pain I felt/feel when I saw/see/felt/feel the death/cutting of those trees is all these things and more. For those trees, far more than I had consciously known, had also found a place inside me – just as I had found a place inside them. Now they are rent from me, rent from this world. There's a hole where life and companionship and the wisdom of centuries once stood. And I live with that loss every day. There it joins all the other griefs and losses that life has brought me, the dead who once loved and held me, the broken dreams, the paths not taken, the open pit mines, the clearcut forests, the polluted streams, the depleted landscapes, the cruelty of human uncaring, of manunkind.

There are millions of us who feel these things. People have been feeling them for centuries. Long ago, the Romantic poets and natural philosophers, the hunter/gatherers, the hand-crafting artisans and farmers who worked in concert with the land warned of what would happen if Earth and Earth people were ground up in the industrial machine, redefined as resources, used to increase the wealth of the few, subsumed into rationalist reductionisms, the living intelligence and soul of animate Earth smothered under oppressive and dismissive monotheisms. The lament of Earth people has sounded for centuries. And for centuries it has been dismissed, denigrated, insulted as foolish, unchristian, naive, irrational, unsophisticated, anthropomorphic, anti-modernist, a denial of progress. Yet here we are, all their warnings manifest, ecosystems in ruins, species dying by the millions, Earth balance spiraling into an unpredictable and uncontrollable future. And millions of us feel it. Every day of our lives.

> *We feel these wounds with the sensitive antenna of our heart's affections for the world. The older we become, the more attuned we are to its touch. And the more attuned we are, the more aware we become of the damage to our home. The more aware we are, the greater the pain that we feel becomes.*

Academics, as slow as they are – even the media (bless their hearts) – have become aware of the Earth Grief that is moving ever more strongly amongst the world's peoples. They're writing articles about it, conducting studies, speaking of an experience they've come to call climategrief/ecologicalgrief/solastalgia. Despite their noticing, there's a problem, a very serious problem, with how they speak of what so many of us are feeling. And because of that problem, very little of what they are doing or writing will be useful for finding the solution we seek or for helping the times in which we live. In fact, *this* particular problem is integral to the destruction of the natural world.

In the beginning, in its earliest form, the problem I'm speaking of is such a subtle thing that it's hard to notice. Each and everyone of us in the united states does it from time to time. (And I've seen it, as well, in the english, the dutch, the irish, the scottish, the koreans, and the japanese, too. Maybe all people do it, I don't know.) But it's ingrained into us, into the cultural system in which we live. It's as automatic as our breathing. It's so natural to us that it's very hard to recognize when we encounter it or do it ourselves. Nevertheless, understanding/noticing/feeling it is the first step in the journey this book is about. To begin to get a handle on it, I will tell you what the problem is. It's dissociating from what you feel.

The comedian George Carlin had a brilliant way of talking about it. He was always incredibly sensitive to the way words are used . . . and how alterations in their use shifted the fabric of the reality in which we live, how those alterations change the way we think, the way we perceive ourselves and the world around us, and also . . . how we act afterward. He knew how dangerous the problem is. In one of his more famous routines he talked about it like this . . .

There is a thing that happens sometimes to men and women who go to war, who are in battle too often, too long, or too intensely. Their nervous systems just can't take it any more. They break down. In World War I they called it *Shell Shock*.

There's an emotional potency in those words, isn't there? Can you feel it as they reverberate inside you? *Shell Shock*. When I hear the words my body and my heart immediately have a sense of what they mean. I tense up, *feel* something, a kind of emotional or nervous system impact that points to the reality of the condition. Each word, as Carlin noted, is one syllable, simple and to the point. Shell Shock. Even the word *sounds* bring some of their reality alive. The softness of *shell* followed by the hard ending of *shock*. *Shell Shock*.

By World War II, though, the name has changed. Now it's *Battle Fatigue*. The immediacy and brutality of war, the impact on the self that *Shell Shock* contains is gone. Battle is still there, so some sort of conflict is happening, but now the soldiers are just tired, as if they need rest or a good nap. Each word is longer, two syllables instead of one. Four syllables in all. The first word is hard, the second soft and fading away (which tends to weaken the impact). Battle Fatigue. Something is being hidden from our gaze – and our feeling self, simply due to the words being used. *Battle Fatigue.*

By the time of the Korean War the same condition becomes *Operational Exhaustion.* The soldiers are still tired, but the hardness and explicitness of *battle* is gone. Operational Exhaustion. The phrase could be about anything... Why did your car stop working? Operational Exhaustion. It's no longer about war or battle and what happens to human beings in war. And the words are longer, eight syllables now. I notice that every time I say it, just by saying it, I move away from my feelings and enter a more emotionless state. A place of mental concepts instead of feelings. There's no sense at all of the horror of war. *Operational Exhaustion.*

During the Vietnam War it changes again. *Post-Traumatic Stress Disorder.* Four words now. Still eight syllables. But the tiredness is gone (which still had something of the body in it – tired, fatigued, exhausted). Now there is little of the body at all. Post-Traumatic Stress Disorder. There's still *stress* but we're no longer exhausted or fatigued, just stressed. That's not so bad, is it? As I speak the words I find myself in some strange place where neither my body nor feelings exist, as if there is no human experience at all. I feel slightly confused. There's just this mental, dissociative concept that I in some strange way exist inside of when I say the words. *Post-Traumatic Stress Disorder.*

Now the phrase has changed again. *PTSD.* Only four syllables now: pee-tee-ess-dee but the dissociation is complete. Finally, every vestige of the human experience is gone. There's no more stress, no contact with the body at all. Just PTSD. The perfect label. There is nothing in it to point to the horror of war, of what happens to human beings in that terrible circumstance, of the shattering that occurs, the damage to their humanness, their lives. There are no feelings in it, it's become completely mental (and it's important to understand that this is the point of the process). *PTSD.*

I was always surprised that George Carlin did not tell us what they called it during the Civil War, for this experience has accompanied human beings in war for as long as human beings and war have been. During the Civil War they called it *Soldier's Heart.*

As I say those words, as they reverberate their meanings inside me, I want to weep. I feel a sadness emerge inside, a tenderness comes as well. My muscles no longer tense – as they do with shell shock. Nor do I dissociate as I do with battle fatigue or operational exhaustion or PTSD. Instead my body softens – *I* soften – and I *feel* more. Something in me breaks and a grief upwells itself. I have a feeling of what those young men, those boys, experienced in that war, of what was broken inside them. *Soldier's Heart* brings with it a compassion that shell shock does not have, nor battle fatigue, nor operational exhaustion, nor PTSD. It reveals something important.

In our time, many of us are suffering *Soldier's Heart*. But rather than talking about it in a way that engages our feeling self, that engages the living texture of our experience, researchers, the media, our friends, our culture, even climate scientists, ecologists, and activists will, for the most part, encourage us to make the dissociative journey from Soldiers's Heart to PTSD. We – all of us – will be, are, encouraged to stand outside the experience and merely talk about it mentally. As if it is over there – something we can point to that is not also inside our bodies and hearts. We are urged, as nearly all people in the west are urged, to behaviorally take on a rationalist dissociation from the world. And to betray, in so doing, our feeling selves.

The reason this is important is that it is our human dissociation from the world that is root to the problems we now face. As James Hillman so poignantly put it, "We have lost the response of the heart to what is presented to the senses." Once our feeling bond with the livingness of the world is broken, once we no longer *feel* the touch of the world upon us, no longer *feel* the response of our hearts to what Earth presents to us every day of our lives, then it becomes much easier for us to be care-less about the damage to our home. Further, once we no longer *feel* the impact of damaged landscapes, the aesthetic wasteland of office buildings and schoolrooms, the brutality hidden inside the civilized language of dissociative terms like *progress* or *anthropomorphism* or *science,* it becomes much easier for us to go along with what is happening. But even worse is the damage it does to us internally, to our own sense of self, to our humanness.

It matters, deeply, the quality of the thoughts (and style of thinking) we take as our own. Some deepen our sense of self, strengthen the best in us, connect us more deeply to truth, to Earth, to our humanness and moral center. Others do quite the opposite. And it is how we feel as we let those thoughts become our own that tells us which is which – specifically, the response of the heart to the touch of those linguistic meanings upon us. And that feeling response leads always to what is behind or underneath those meanings – the psychological and epistemological orientation of the one or ones who composed them. Thus, to allow the thoughts of others to take up residence inside us without a simultaneous discernment of their character or analysis of their nature is often unwise. As William Gass once put it . . .

> *For what is it to take a guest of this kind into the interior of the soul, from whence words rise like a sudden spring; what is it to offer your hospitality to the opinions and passions, the rhythms and rhetoric, of another, perhaps far from perfect, character and mind? . . . In societies which depend principally upon the spoken word to establish and maintain community, the real origin of one's words is a serious, even critical matter. . . . Rhetoric in the abstract [is composed of] words quite free of responsibility to anyone. It is no wonder that Socrates feels uneasy in their presence.*

It is our feeling response, that is, *the mood or atmosphere or climate of mind that arises in us as we read*, that tells us the true nature, and epistemological value, of the thoughts of others as they enter inside us.

When I take on dissociative thinking as my own – let its thoughts become my thoughts, allow myself to see the world through its eyes – the further away I become from my own feelings, my body, my interbeing with the world. The further I am from myself. I begin to doubt my own experience, feel a pressure to apologize for what I feel, start to distance myself from my heart's responses to the touch of the world upon me, to conform to a different way of being and thinking. I lose my moral center. And over the decades of my life I have found that there is a soul damage that comes from that, from acquiescing to the dissociation. And I can't escape the feeling that somehow, when I do so, I have started to collaborate in my own, and Earth's, oppression.

This dissociative process is widespread; it's incredibly common in daily life. Regrettably it is also common among environmentalists, activists, ecologists, scientists, psychologists, physicians. In fact it is integral to the western culture in which so many of us live.

The dissociation I am speaking of presents itself immediately when the terms commonly used to talk about Earth grief are used. *Climate grief. Ecological grief. Solastalgia.* It was perhaps Einstein who said that if a scientist could not explain what he knew in such a way that a child could understand it then he didn't really understand what he did – or the nature of his own work. There is something crucial and important in that observation, something each of us would do well to remember every day of our lives.

Every child understands the word *Earth* and each and every one of them understands sadness, which, in many ways, is only a milder form of *grief*. It's something all of us feel from time to time in our lives. But *climate, ecological, solastalgia* . . . each one is a dissociative term. Do any of them have the power of *Shell Shock* or *Soldier's Heart*? Do they ground any of us in our bodies or our feelings? What do they actually mean? None are easily explainable to a child – or to the part of us that feels, that still is a child in the best sense of that word. You can point to Earth, you can't point to climate or ecological, and as for solastalgia, well, *gesundheit*.

They are grown-up words, adult words, clever words that professionals use. They are many-syllabled, which is often an indication that something is wrong . . . especially when talking about something as important as that which we now face – and how we *feel* about it. Each of the terms dissociates the speaker *and* the hearer from grounding in their body and moves them upward into the mind, away from feelings. They are *thinking* words, not feeling words.

Even in well-meant attempts to address the problem, the problem is interwoven. And there is a progression to it, just as there is with words describing the condition that those in the nineteenth century called *Soldier's Heart*. It travels from the experience itself at one end of the spectrum to a complete and utter dissociation at the other. Here is what that movement feels/looks like among those of us talking about Earth grief.

> *It's very hard for me to deal with what's happening. I'll be driving along and then suddenly come upon a shattered landscape. The trees I loved as a child are gone and the whole place has changed. It's like someone tore a hole out of the world . . . and out of me. I keep thinking I shouldn't feel it so strongly but I do and I feel like I can't tell anyone about it because no one seems to feel about it the way I do. So I keep it to myself. And the pain and the grief just builds and builds and builds. There are days when it's so bad that I think I just can't go on.*

That is the experience itself, told simply, without dissociation. But what most people do instead is this:

> *It's hard to deal with what's happening. You're just driving along and then suddenly come upon a shattered landscape. The trees you loved as a kid are just gone, the whole place has changed. It's like someone tore a hole out of the world . . . and out of your heart. You think maybe you shouldn't feel it so strongly but you do and you can't tell anyone about it because no one seems to feel about it the way you do. So you keep it to yourself. And the pain and the grief just builds and builds and builds. Some days you know, you feel like you just can't go on.*

It's a subtle difference but in it lies the destruction of our world.

Most people in the western world talk about their experience of life and self using the kind of sentence structures in that second paragraph and they do it a lot of the time. They talk in what is called *second person singular* rather than first person singular. It occurs so often, is so much a part of our culture, that it doesn't sound odd to the ear. But what second person singular does is distance a person from their feelings. It points to the feelings but the feelings now are slightly outside the self, there is no ownership of them, no responsibility for them. The pain is more bearable (and this is why it's sometimes healthy to do it – just to get a little distance, a break) but it also decreases the vulnerability of the self. With second person singular we do not stand naked in our pain, our self revealed to the gaze of others. It allows us to remain partially clothed while pointing to the fact that nakedness exists out there someplace. It also alleviates the social discomfort others commonly feel in the presence of real feelings, of feelings that go deep, that touch upon a person's interior world. Western cultures are deeply uncomfortable with nakedness... of whatever sort.

Do you get a sense of the difference in these two ways of speaking? Try it yourself and see how you feel when you do. Imagine yourself saying the following sentence to someone you have just met and are talking with: *I felt really scared when it happened.*

That feels uncomfortable, perhaps a bit frightening, doesn't it?

Now imagine saying it this way: *You know how it is, at moments like that you feel really scared.*

Not so uncomfortable or scary, is it? It's much less vulnerable. Much less intimate.

I notice that when I choose to speak in first person singular, when I say "I," the feelings are stronger, more immediate. I am *in* the experience and what's more, the simple act of being in the experience unlocks something inside me, the door to grief and pain and fear opens. Something real enters the room then, something that was not there when I used "you" instead of "I."

It took me a long time to learn that one of the most important things of all on the journey through grief is to *feel* what I am feeling, to not dissociate from it, to be real, to be genuine, to be the self that is most centrally me as much as I possibly can, every minute of the day, in everything I do, every conversation I have, every word I use.

Using the word "I," taking on and owning the feelings without dissociation, opens the doorway that allows the journey into and through grief to be taken. (It is also frightening. Every time you do it you will be breaking a cultural injunction. Every time you do it you will become somewhat naked in public – and perhaps more difficult, naked to yourself. And you will often feel deeply uncomfortable, even afraid. It takes time to get used to it.)

Still, it's the only way I know to get to what can be found on the other side of this grief, to understand the messages, meanings, and communications that are woven into the grief Earth is sending us. It's the only way I've found to get to the resolution of the problem and the pain that Earth destruction brings. Dissociation closes that opening – which is why people do it.

One of the hardest things there is, is to just feel the pain and grief that surrounds us now, to acknowledge it out loud, to be inside the aloneness of it, to not blame anyone for it, to just feel it simply because it is in you to feel it. It's hard but I think it is time to make a different choice, to be braver as a people than we have been, to take the journey that the grief of Earth is asking of us.

> *What we are so often exhorted to do, what we so often want to do, in response to Earth grief is to avert our eyes, to move our seeing into and onto imaginary and optimistic landscapes of what* **can** *be, to divert the pain into some possible future. But in so doing we become intentionally blind, retreat as best we can into a childhood state of innocence, step inside an illusory safety so we won't feel the pain that is waiting for us there, in the reality of what the powerful and dissociated have done to Earth, in the place where it's possible to see the true face of the* **now** *in which we live.*

I know all this. Yet I still catch myself dissociating in all the ways I have been taught to do. It's a *very* ingrained pattern. So I work hard, right there in that minute, to, as quickly as I can, alter what I am saying, correct myself, and use the word "I." My heart opens when I do, becomes more tender, more real, because I'm no longer stuffing feelings into a box so that I won't feel them as strongly. In an important sense, as John Seed once described it, I reinhabit my interbeing with the world. But more crucially, as e. e. cummings puts it, I become myself . . .

> "[Feeling] may sound easy. It isn't. A lot of people think or believe or know they feel – but that's thinking or believing or knowing; not feeling. . . . Almost anybody can learn to think or believe or know, but not a single human being can be taught to feel. Why? Because whenever you think or you believe or you know, you're a lot of other people but the moment you feel, you're nobody-but-yourself."

The place to begin is with the simplest thing of all: how you feel right now this very second. Like this . . . *I feel* (fill in the blank – mad, sad, glad, scared, for instance) *because* (fill in the blank) *and what I need right now is* (fill in the blank).

Like this: I feel pain and grief and emptiness because the trees I loved have been cut down and what I need right now is for you to hear my pain and grief and emptiness and for you, the one who loves me, to hold me while I grieve, to hold me as the pain comes out of me in these sounds, in this sea that flows from my eyes, until I have let enough of the pain out that I can deal with the grief I feel, until I can go on.

Dissociation travels further from this simple beginning, of course. As with the journey from *Soldier's Heart* to *PTSD*, it becomes ever more extreme in the process. And despite its apparent reasonableness, it becomes increasingly dysfunctional and pathological the further it goes. So, let's look at the dissociative journey it takes and how it often feels/looks on the way. (I have scores of examples and my usual tendency is to put in ten or twenty to overwhelm with the point but I won't, I'll just include three.) To begin with, here is Melissa Harrison's article "Feeling severely distressed about the climate crisis? You're suffering from solastalgia," in *The New Statesmen*, online . . .

> *Have you cried yet? No? Don't worry, you will. Maybe it will be a news story about the last ice in a glacier, the last living coral on a reef, or the extinction of a bird you have never seen*

> *and never will. Perhaps a TV report, or even a tweet, will finally make you weep: one about deforestation, wildfires and the scorched and flooded world your children's children will inherit.*

That's a media article, rather flippant, and quite dissociated. It doesn't feel very good, does it? It doesn't really do anything to help me with the grief I am feeling. Quite the opposite. I feel talked down to, patronized, alone, uncompanioned, my feelings of grief superficialized. The style of writing allows the writer to remain dissociated, to keep herself concealed inside the words and the story she's telling. It allows her to remain safe.

The dissociation's far more extreme when we get to Wendy Shaw and Alastair Bonnett in their article "Environmental crisis, narcissism, and the work of grief" in the journal *Cultural Geographies* . . .

> *This sense of detachment, accompanied with the overwhelming experience of environmental change, occasions a variety of grief reactions. Collectively, we term this the work of grief. As noted earlier, this idea may be contrasted with the traditional and influential model of "grief work," which assumes distinct and discrete stages of grief. Indeed, in* Living in End Times, *Zizek (2010) posits that humanity has begun to grieve as global capitalism reaches its inevitable terminal crisis and proceeds to identify stages in the grieving process, with denial the first. . . . Zizek's cultural references return us to our identification of a relationship between the work of grief and narcissism. The point can be developed by revisiting some of the examples already provided by analysts of the psycho-social consequences of environmental crisis.*

How do you feel *now*?

To clarify: I am not just talking about emotions here – mad, sad, glad, scared (though these are always important to identify when language such as this takes up residence within us) – I am also talking about the *atmosphere* or *mood* of the thing, what Tim Leduc identifies as its *climate of mind*, what William Gass calls the *secret kinesis of things*. Like when you go into a restaurant with a friend and suddenly stop and turn to each other and say, "This place feels weird, let's leave." And the atmosphere or mood that arises? If you stay with it, contemplate it, observe what is happening inside you in response to what you have taken inside you, you will find that specific states of mind and being have arisen because of it. There is a nearly automatic associational experience that occurs – certain ways of thinking, feeling, and being arise in response. This is deeper and more pervasive than feeling sad in response to a tragic story. It is the climate of mind from which the story has been told. So ask yourself, while Shaw's and Bonnet's words are still reverberating within you . . .

Do you feel more or less in touch with Earth, your body, the grief that Earth devastation brings to the heart? Do you feel more alive or less? More whole or less? More capable of dealing with your feelings of Earth grief or less? More the self you centrally are or less?

Rather than getting caught up in the mental communication of the words and dissociating from the response of your heart to what you are reading, what is the primary feeling and state of mind the article engenders in you? Disregard what it is saying, *how do you feel?* (I feel lonely and sad, dissociated, a bit empty, a bit lost from myself – this tells me a great deal about the article itself and the psychological and epistemological orientations of the people who wrote it.)

The Shaw and Bonnet article, and the feeling it engenders in the reader, is a perfect, though hardly unique, example of the dissociation that is common in the sciences, where feelings and genuineness are suspect and, when found, are rooted out with grim insistence. The feeling of wrongness it has, the impact it makes on the aesthetic dimensions of those who read it, the responses of the heart to its underlying orientations and climate of mind are far more important than anything said in the article itself.

Dissociation can always get worse, of course. And it does so in Thomas Doherty's and Susan Clayton's extensive article in *American Psychologist* . . .

> *There are numerous accounts of subclinical depressive emotions, guilt, and despair associated with climate change and other*

global environmental issues. Fritze et al (2008) discussed how "at the deepest level, the debate about the consequences of climate change gives rise to profound questions about the long-term sustainability of human life and the Earth's environment." In this vein, Kidner (2007) described the loss of security engendered by uncertainty about the health and continuity of the earth's natural systems and how the impact of these emotions tends to be underappreciated because of a lack of recognition of subjective feelings of environmental loss in traditional scientific and economic frameworks.

It's ironic that they comment that *"these emotions tends to be underappreciated because of a lack of recognition of subjective feelings of environmental loss in traditional scientific and economic frameworks"* while writing using a non-emotional, dissociative tone and structure. Their article, and way of thought, is in fact a very good example of the problem they are talking about.

Despite our culture's overall disavowal of the importance of the feeling dimension of life, what we feel as another's thoughts or an external environment takes its place inside us is always important. How we feel in any particular moment is a clue to the deeper meanings inside the environment (or words) that are touching us. It is a clue to their aesthetic dimensions.

Just as the feeling we have when we enter a clearcut forest tells us about the aesthetic dimension of the forest at that moment in time, so, too, does the feeling that emerges when reading a text tell us of *its* aesthetics: whether we feel more alive or less, more in touch with our humanness or less, more in touch with our capacity to love or less, more whole or more fragmented, more uplifted or downcast, more in touch with the best in us or more in touch with the worst. (In truth, none of these are dependent on the topic, only on the aesthetic dimension inside the words and sentence structures that are being used, which themselves are generated out of the psychological/epistemological orientation of the writer.)

What a person feels when they unexpectedly come upon a human-damaged forest is a combination of two things: the kind of wound (and its ramifications) that touches them **and** the particular climate of mind of the people who destroyed the wholeness that was once present.

> *Those who clearcut a forest are able to do so in the way they do because they live within a particular climate of mind – one that has dissociated itself from the livingness of the world. And yes, there are other ways to harvest trees, ways that do not possess or come from a pathological climate of mind.*

That particular climate of mind is integral to every form of ecological damage we encounter. It is now *in* Earth itself in so many locations it cannot be escaped. We are immersed in, surrounded by, such wounds . . . *and* the climate of mind that produced them.

When the aesthetic dimension of a thing is compromised or damaged a particular *feeling* emerges from the heart. There's a *wrongness* to what has **happened** and that wrongness has a particular, unique feeling dimension to it. And there's a great deal of information *inside* the mood or atmosphere that arises. Contemplation of its textures will reveal that information, give a far greater understanding of what has happened than any reductive analysis can or ever will. It will tell you about the nature and form of the complex ecological disruptions that have come from the damage . . . and their wider implications and effects.

When you open to the touch of the world upon you as a habit of mind and heart, when you care about a landscape that has been disrupted, *feel* the atmosphere, mood, climate of mind and heart that it is giving off – that is, when you *love* it – it is possible to perceive a great many more things than if you do not. It's just the same as it is with your beloved, the one who means more to you than any other person you will ever know. Because you love, you *see*, you *feel*, you *perceive*, and you *know* in far deeper and more subtle ways than you would otherwise.

This capacity, it is important to understand, is an evolutionary innovation of very long standing. It is integral to our survival for it allows us to determine the qualities or meanings that surround us and which we encounter daily. At its simplest, it allows us to identify dangerous or threatening circumstances even if we have no rational basis for doing so. But when trained as insistently and for as long as we train the rational mind it can parse encountered meanings with great sophistication and subtlety.

Thus the perceptual analysis of an environmental wound will be far more sophisticated and accurate if the feeling response to a clearcut forest is immediately compared with that of a healthy forest – for it is by comparison, and an analysis of differences, that we learn to distinguish the subtle characteristics of things.

Healthy forests generate a particular, unique feeling in the heart, initiate a particular climate of mind inside us, just as clearcut forests do. There is a certain texture, a specific aesthetic dimension, that they possess. (And this will vary in its expression due to the location-specific form each healthy forest has. The landscape in which it has emerged shapes its form and nature.) In healthy forests there is depth, a richness, a feeling of healthiness, of happily engaged life. Such forests have a climate or property of mind that can be felt, that *is* felt whenever they are encountered. And that climate of mind is one that human beings can take on as their own – just as we take on the thoughts of others when we read the words that they write.

Forests are useful for many things but perhaps most important is that they are good to think. Taking on a healthy, old-growth forest's climate of mind allows the mind to think and the eyes to see and the heart to feel in a very different way than we have been taught to do in our schooling, our families, our cultures. And the particular climate of mind that arises when we do is essential to our successful habitation of Earth. It is core to the different path this book is about.

Clearcut forests, and the climate of mind within them, are very different, of course. When *that* climate of mind is taken on as one's own, as the mind thinks through it, as the eyes see through it, a very different way of being, and a unique state of mind and heart, emerges. And it is a pathological one. It leads to the state of being and world our species now finds itself in. If you practice taking on the climate of mind that is held in a damaged forest you will find that dissociation emerges automatically; it's integral to that climate of mind. The feelings of your heart, and your responses to the touch of the world upon you, will be subdued, even repressed. A form of alienation occurs accompanied by the loss of a healthy sense of self.

There's a series of questions that immediately arise for me when I engage in a comparison between a clearcut and a healthy forest, as I take on those two very different states of mind and being, when I allow myself to be subsumed by them. They are: What happens to this Earth upon which our existence depends when so much ecosystem damage occurs that the feeling a clearcut forest generates becomes predominant worldwide? What happens to us, the human species, individually and collectively, when we continually live within and are surrounded by that kind of feeling, that climate of mind and spirit? What happens to the kindred species with which we share this planet? What happens to our children? And shouldn't these things be considered before even one forest is clearcut? Shouldn't we understand the long term implications of that kind of behavior before it becomes a way of life for our species?

These deeper dimensions, these aesthetics of mind and heart I am talking about, can't be found through the intellect, for the intellect, as good as it is for many things, is not the part of us that senses meaning. Meaning is found through our heart's response to what is presented to the senses. It's a *feeling* thing, not a thinking thing. Thinking comes afterwards, and part of what thinking is good for is allowing us to capture in language the subtleties inside the aesthetic events we encounter.

Through contemplation of the feeling state that emerges from contact with a healthy forest it is possible to perceive the complex communications that occur between plant species in that forest, between the trees, between the plant community and the many forms of life that live there. It is possible to understand the functions that particular forest is fulfilling in the region in which it is embedded. It is possible to feel, *immediately*, from the impact upon the heart, what is being lost when all that is destroyed and what the likely longterm impacts will be . . . on Earth, on humans, on future generations, on every kind of life. Again, it is in the contrast between the way healthy and damaged ecosystems feel that the feeling sense can be trained, over time, to distinguish subtleties of the informational complex that is held within them.

All this is true as well of any written material that is read. But what is revealed when the feeling state generated by an article is contemplated is the climate of mind of the people who wrote it, the beliefs that underlie what they are saying, the psychological orientations the writers possess. As with a healthy or a damaged forest, the feeling you have when you enter the world of the article will tell you of its aesthetic dimensions. It will reveal how healthy or unhealthy it is, damaged or whole, as well as the psychological orientations, beliefs, and climate of mind of the writers. If the aesthetic dimension inside the article is damaged – which most often occurs because of a particular psychological wound or distortion in the writer – there will be that same sense of wrongness (even if in lesser degree) that is felt in a damaged ecosystem or forest. *And that is the important thing, that feeling.* Whether or not the intentions of the writer or writers is "good" is irrelevant. What is relevant is the outcome of their behavior, the ripples their work sends outward into the world. The type of corn they are growing. Importantly, the climate of mind that underlies, or is embedded throughout, the article itself is taken on, to one degree or another, by everyone who reads that article. They carry it within them afterward and it has an impact on who they are as a person and on how they behave and feel about themselves and the world around them. As Gass has said, "The real origin of one's words is a serious, even critical matter."

> *To be clear here, I don't necessarily avoid either clearcut forests or dissociated articles. But when I go into them I **know** what I am entering. I have spent years immersing myself in such things, just to see what happens when I do. And I've learned a lot from doing so. If we cannot immerse ourselves in the climate of mind that is causing so many of our troubles, how can we understand its nature, its underpinnings, its elements? The only way to respond to it with the depth that is necessary is if we do understand it and understand it well.*

As I contemplate the feeling dimension of these articles (and the climate of mind they generate within me), more questions arise: What will be the impact on the ecological structures of the planet if dissociated, feeling-less mentation becomes dominant throughout the world? If Earth is continually seen through that lens? If science takes that on as its default state of mind? If all scientific research is conducted from that orientation? If people are trained to be disconnected from the ground upon which they walk, the life forms that surround them, even each other? What happens when millions of people are trained out of their capacity to feel the responses of their heart to the touch of the world upon them? What happens when this becomes our *reality*?

I have been thinking about these questions for half a century now. The answers, it turns out, were and are all around me; I see them every day of my life in the instability of our world. But in the end, I have found that the most important question for me is this: Am I willing to be complicit in that process? And the only answer that I can honestly give is: *No, I am not.* And so I work to remain aware, to ask always the first and most important question about everything I encounter: How do I feel *right now*? After I have analyzed the climate of mind I've encountered, to whatever depth or complexity I've chosen in that moment in time, I then decide how to respond to it.

I cannot, by myself, end the dissociative climate of mind that has taken over the western world, but I can decide, in every moment, what kind of human being I will be as I live in the midst of it. And I can decide how I am going to respond, in general as well as to any specific manifestation of it, inside me, in the secret chambers of my own heart, as well as out there, in the wildness of the world.

Making my way through scores of journal articles and media reports similar to these has been a weary task. My grief became more pronounced, my stomach hurt (always a bad sign), and if I really allowed those words to become my words, to live in my mind – if I allowed myself to see the world through their lens – I felt dissociated from my body, lost from myself, from my feelings, from any sense of the right, from any feeling of land ethic.

The first step in the decolonization of our minds and hearts is recognizing these kinds of impacts on the self and doing something quite different than going along with them as if they are somehow legitimate. Specifically, rather than allowing such a climate of mind to unquestioningly become *my* climate of mind, what I have learned to do is to become an experiencer of the experience as it is happening. That is, I allow myself to take it on and then watch and feel what happens inside me as I do. I become my own epistemological laboratory; I watch how my identity and sense of self alters under the impact of the climates of mind I encounter. And in so doing, I learn many things. About myself, about others. About our world. About the alterations that belief systems create inside me as I take them on, and inside others as they take them on. And I can then choose what I want to do and be. Foremost among those choices is the abandonment of epistemic colonization.

The first act of disobedience is to feel again in all its complexity – to let Earth touch the self and tell of its life through the feelings it engenders in our hearts. (This is the moment where rationalists begin to use words like *anthropomorphism*, to insist that their *mechanomorphist* reality is fundamental to the legitimacy of all personal and public thought and comment. It is the moment when the epistemological mistake Gregory Bateson talks of becomes aggressively engaged.) The second act of disobedience is contemplating what has been felt. The third is taking a different path, *literally* becoming the change we want to make.

Unfortunately, the kind of speech found in the few examples I have included here is common in nearly all media reports and in pretty much all the academic journals. (There is an extensive list in the bibliography of this book if you wish to deepen your experience of them.) Exceptions are extremely rare. (The ones that are, I have marked with an asterisk; they helped me a great deal on this journey. They might help you as well.)

Purportedly the writers (especially the academics) are deeply concerned about the impact of Earth devastation on each and every one of us, the emotions and feelings we are living with every day, and how hard it is to continue our lives as those feelings/emotions increase in intensity over time, as the devastation gets worse, as the powerful do nothing to stop what is happening. But the truth is, nearly all the writers are dissociated to one degree or another. They talk about Earth grief but it's not possible to *feel* it in the articles they write. (Thankfully, that this is a serious problem is beginning to be recognized: "Reporters covering the climate crisis," the writer Kyle Pope has said, "must be more than stenographers of tragedy.") The dissociation, especially in academic articles, is almost always total and complete.

But now . . . read the passages that follow. This is what true grief *feels* like when it touches you, when it fills the words someone writes. It is the real and genuine brought naked and unconcealed into language. And it is this feeling *and only this feeling* that leads to what can be found on the other side of grief. It is the key to the journey that Earth and our own sense of the right is asking us to make.

These words come from various sections of the book *Time Lived Without Its Flow* by Denise Riley, which is itself abstracted from the journals she wrote after the death of her son Jake . . .

> *In these first few days I see how rapidly the surface of the world, like a sheet of water that's briefly agitated, will close again silently and smoothly over a death. His, everyone's, mine. I see, as if I am myself dead.*

> *Apparently almost half a year has gone by since J disappeared, and it could be five minutes or half a century, I don't know which. There is so very little movement. At first I had to lie down flat for an hour each afternoon, because of feeling crushed as if by a leaden sheet, but by now I don't need to lie down. This slight physical change is my only intimation of time.*

> *[I have] a strong impression that I've been torn off, brittle as any dry autumn leaf, liable to be blown onto the tracks in the underground station, or to crumble as someone brushes by me in this public world where people rush about loudly, with their astonishing confidence. Each one of them a candidate for sudden death, and so helplessly vulnerable.*

> *Wandering around in an empty plain, as if an enormous drained landscape lying behind your eyes had turned itself outward. Or you find yourself camped on a threshold between inside and out. The slight contact of your senses with the outer world, your interior only thinly separated from it, like a membrane resonating on a verge between silence and noise. If it were to tear through, there's so little behind your skin that you would fall out towards that side of sheer exteriority. Far from taking refuge deeply inside yourself, there is no longer any inside, and you have become only outward. As a friend, who'd survived the suicide of the person closest to her, says: "I was my two eyes set burning in my skull. Behind them was only vacancy."*
>
> *This state is physically raw, and has nothing whatever to do with thinking sad thoughts or with "mourning." It thuds into you. Inexorable carnal knowledge.*
>
> *Now I've no sense of any onward temporal opening, but stay lodged in the present, wandering over some vast saucer-like incline of land, some dreary wide plain like the banks of the river Lethe, I suppose. His sudden death has dropped like a guillotine blade to slice through my old expectation that my days would stream onwards into my coming life.*

That is grief. *That* is how it feels. And one of the terrible tragedies of our time is that in nearly all of the journals and articles about Earth grief, grief itself does not make an appearance. The authors point to it but they write as if they themselves are not a part of this scenario we call Earth, as if they can step back from it, as if these feelings can be talked about dispassionately, objectively, as if in fact they themselves don't feel them. Very, very few of them have made the journey into and through that grief; very few of them have anything to tell us about grief that is useful. Worse, that stepping back, that disowning of the feeling self, is itself one of the primary generators of the ecological devastation we can no longer ignore. The *way* they approach Earth grief *is* the underlying problem.

To endure as a species, our language must become an integral expression of the ecological realities of this planet so that merely by speaking our conscious awareness is more fully embedded within those ecological realities. And this includes, always and most importantly, how we feel. The more dissociated our language becomes, the more we, and our species, are readying ourselves to fall.

The grief that we feel is intended to teach us something, something crucial to our habitation of this world. Inside it are many messages and the only way to find them is to journey into and through the grief itself, to wherever it takes us, to enter its world, to sit at its feet and learn what it has to teach us.

There is much more that needs to be said about Earth grief and its teachings (and I will get to it in a little while) but first we need to talk about shame and guilt.

INTERLUDE TWO

"Once You Are Real You Can't Become Unreal Again. It Lasts For Always."

Grief is a resonating chamber for an unknown part of the personalty.

ROBERT BLY

Anything will give up its secrets if you love it enough.

GEORGE WASHINGTON CARVER

My best friend lives inside me – though it took decades for me to realize it. He's the mature, now-70-year-old, fully alive child that stands next to the tenderness of my heart. My heart, though it has more life experience now, is still the same heart that the very young, four-year-old child inside me knew as he sat under trees talking to flowers and stones. It is the same heart that fell in love with wilderness when I was young, that heard the silver speech of streams caressing stones as they moved toward the sea, that discovered the shadowed mystery of lichen-shaggy forests and therein found something essential to the soul of me. The same heart that loved my grandmother to distraction, that lay at night in my great-grandfather's arms as he told me stories of his youth and what the world was like in the last decades of the nineteenth century.

It is the same heart I was encouraged to abandon as I grew older, as I was schooled, as I rubbed up against the sharp edges of strangers, as I began to sell my body for money rebuilding houses and restoring broken homes. And it is the same heart that I found years later, in the rubble of all that abandonment, pain, and fear. The same heart that I slowly brought forth again from its imprisonment, the same heart through which I learned to feel and love again.

Someone who understood these things was Margery Williams, who wrote *The Velveteen Rabbit* long ago.

> *For a long time he lived in the toy cupboard or on the nursery floor, and no one thought very much about him. He was naturally shy, and being only made of velveteen, some of the more expensive toys quite snubbed him. The mechanical toys were very superior, and looked down upon everyone else; they were full of modern ideas, and pretended they were real. The model boat, who had lived through two seasons, and lost most of his paint, caught the tone from them and never missed an opportunity of referring to his rigging in technical terms. The Rabbit could not claim to be a model of anything, for he didn't know that real rabbits existed; he thought they were all stuffed with sawdust like himself, and he understood that sawdust was quite out-of-date and should never be mentioned in modern circles. Even Timothy, the jointed wooden lion, who was made by disabled soldiers, and should have had broader views, put on airs and pretended he was connected with Government. Between them all the poor little Rabbit was made to feel himself very insignificant and commonplace, and the only person who was kind to him at all was the Skin Horse.*
>
> *The Skin Horse had lived longer in the nursery than any of the others. He was so old that his brown coat was bald in patches and showed the seams underneath, and most of the hairs in his tail had been pulled out to string bead necklaces. He was wise, for he had seen a long succession of mechanical toys arrive to boast and swagger, and by-and-by break their mainsprings and pass away, and he knew that they were only toys, and would never turn into anything else. For nursery magic is very strange and wonderful, and only those playthings that are old and wise and experienced like the Skin Horse understand all about it.*

> "What is REAL?" asked the Rabbit one day, when they were lying side by side near the nursery fender, before Nana came to tidy the room. "Does it mean having things that buzz inside you and a stick-out handle?"
>
> "Real isn't how you are made," said the Skin Horse. "It's a thing that happens to you. When a child loves you for a long, long time, not just to play with but REALLY loves you, then you become Real."
>
> "Does it hurt?" asked the Rabbit.
>
> "Sometimes," said the Skin Horse, for he was always truthful.
>
> "Does it happen all at once, like being wound up," he asked, "or bit by bit?"
>
> "It doesn't happen all at once," said the Skin Horse. "You become. It takes a long time. That's why it doesn't happen often to people who break easily, or have sharp edges, or who have to be carefully kept. Generally, by the time you are Real, most of your hair has been loved off, and your eyes drop out and you get loose in the joints and very shabby. But these things don't matter at all, because once you are Real you can't be ugly, except to people who don't understand. . . . "Once you are Real you can't become unreal again. It lasts for always."

Every so often, during the years of my teaching, I would ask my students to reach out and touch the earth closest to them. It is rare that any of them reached up and touched their own face. You are Earth expressed into human form. All of us are.

What have you done with the garden that's been entrusted to you?

CHAPTER THREE

Shame and Guilt

Unlike guilt, which is the feeling of doing something wrong, shame is the feeling of being something wrong.
                                        MARILYN SORENSEN

Shame is the lie someone told you about yourself.
                                        ANAIS NIN

I don't believe in collective guilt. The children of killers are not killers, but children.
                                        ELIE WIESEL

One of the terrible difficulties for many of us who feel Earth grief is our feelings of shame and guilt. They come from a sense of responsibility for what is happening to Earth, as if you and I and every other human being are to blame. That, after all, is what the term *the anthropocene* means: the geological era of human-caused climate change. And comments about human responsibility are seemingly everywhere; if you read the news you have seen them, probably more often than you wish . . . and, as well, felt the impact of the words upon you. I have. I encounter them every day of my reading life.

Here is one from Alan Betts, an atmospheric researcher.

> *In just a few centuries, with the advent of science and technology, powered by fossil fuels, humanity has moved from feeling largely at the mercy of the Creation to a world-view where we*

> *thought we were omnipotent – that our power was limitless. We have searched the Earth for all its resources, and let our population and consumer societies expand without limit; while dumping out wastes and polluting land, oceans and the atmosphere.*

And one from Stephen Emmott at *The Guardian* in his article "Humans: the real threat to life on Earth" . . .

> *Earth is home to millions of species. Just one dominates it. Us. Our cleverness, our inventiveness and our activities have modified almost every part of the planet. In fact, we are having a profound impact on it. Indeed, our cleverness, our inventiveness and our activities are now the drivers of every global problem we face.*

And Donald Spady in *Pediatrics and Child Health*, volume 14, number 5, May/June 2005 . . .

> *Every child, everywhere, all three billion of them over the next 50 years, will have to confront unprecedented and generally adverse changes within their societies during their lifetime. Just as our economy is undergoing incredible change, so is the environment. Just as today's economic changes are distressing to most, so will be tomorrow's environmental changes. Just as we precipitated these economic changes because of our arrogance, greed and hubris, so have we caused many of the environmental changes.*

And finally, Poonam Ghimire at *Voices of Youth* in her article "My Earth, My Responsibility" . . .

> *Despite unavoidable free services provided by the Earth to humans, we are not able to pay off her kindness to us. Rather we humans are being cruel to our Earth with selfish activities. . . . Every day we produce tons of degradable and non-degradable waste, and throw it anywhere recklessly. Smoke and harmful*

gases from our homes, vehicles and industries are suffocating her. We are disposing of dirty sewage, drainage and even chemicals recklessly, although we know that more than 7 billion humans, along with all plants and animals in this world, rely on water for their lives.

There are hundreds more articles like this, all of them using the words "we" and "our." As with many (though not all) universals, there are significant problems in the usage and because of it the articles themselves. Specifically, each contains a terrible logical fallacy that has, as all logical fallacies do, serious real world repercussions. In other words, the widespread sense of human responsibility that these articles foster becomes for many people feelings of shame and guilt – and even, to a certain extent, a kind of self-hatred – simply for being a human being, for contributing to the devastation of Earth.

Here is what that looks like, from an Ash Sanders article in *The Believer* magazine, where she talks about her friend Chris and his struggles.

> *He harvested mesquite in a grove of trees and picked wild radishes and mallow in a nearby field. . . . Chris harvested only fallen fruit – he felt this was less invasive than picking from trees, and his aim was to tread lightly on the earth, to be almost invisible, in order to cause as little harm as possible. . . . "I couldn't accept the privileges of humanity when I didn't want any part of humanity," he told me. Eating fallen fruit and sleeping outside, however, didn't provide him relief from his feelings of guilt and foreboding. He began to feel a dread that was inescapable and all-consuming. A devastating depression that he had suffered a few years before that fall semester returned. Normally a math phenom, Chris started failing his tests. In his apartment, he would sit in the dark – he didn't want to waste electricity – to listen to records and cry. "I felt like I was slowly dying," he said. . . .*
>
> *At home, Chris's family had a hard time understanding the intensity of the self-denial that governed his life. His father and*

> sister blamed his breakdown on abuse that Chris had suffered as a child; they believed his desire to escape society was a projection, an act of taking responsibility for something that wasn't his fault. But Chris had a different explanation. When he was fifteen his father had taken him and his sister on a trip to Mount St. Helens. Halfway up the mountain, they had passed clear-cut land. As Chris recalls, one moment there was only evergreen forest and the next moment there was nothing – just bare ground and stumps as far as he could see. A word came into his mind: evil. From that day forward, something shifted in him. He didn't want any part of such destruction. . . . He was offended by his family's attempts to find explanations in his psychology for the problems he thought of as external to him. "Why does my grief have to be because something happened to me?" he told me. "They made it sound like I had a psychosis or a mental breakdown and that this is just the form it took, when really, shouldn't anyone who is ethical and compassionate also choose to opt out of society?"

Here's another example by Mary Annaise Heglar, from an article at *Vox* . . .

> I'm at my friend's birthday party when an all-too-familiar conversation unfolds. I introduce myself to the man to my left, tell him that I work in the environmental field, and his face freezes in terror. Our handshake goes limp.
>
> "You're gonna hate me," he mutters sheepishly, his voice barely audible over the clanging silverware.
>
> I knew what was coming. He regaled me with a laundry list of environmental mistakes from just that day: He'd ordered lunch and it came in plastic containers; he'd eaten meat and was about to order it again; he'd even taken a cab to this very party.
>
> I could hear the shame in his voice. . . . Sadly, I get this reaction a lot. One word about my five years at the Natural Resources Defense Council, or my work in the climate justice

movement broadly, and I'm bombarded with pious admissions of environmental transgressions or nihilistic throwing up of hands. One extreme or the other.

And finally, Jennifer Jacquet from her article in *Lapham's Quarterly*, "Human Error: Survivor Guilt in the Anthropocene," where she talks about and explores some of the ramifications of the death of the last of the pinta island giant tortoises, the one they had named Lonesome George...

> *I wished the museum staff had said something about how difficult it is to be a member of the species that bears responsibility for the Pinta tortoise's demise. . . . Dealing with the disaster we have created means finding a way to reckon with our guilt for causing it. "Why stick around to see the last beautiful wild places getting ruined, and to hate my own species, and to feel that I, too, in my small way, was one of the guilty ruiners?" asked Jonathan Franzen in 2006. "The guilt of knowing what human beings have done" is how conservation biologist George Schaller described the feeling he gets when he looks at the Serengeti. In 2008 Schaller made one of the most definitive statements of Anthropocene-inspired self-reproach. "Obviously," he said, "humans are evolution's greatest mistake."*

The use of the universal "we" and "our" in the context of climate change is not uncommon in the media, in activist literature, in books or in texts on ecology. But they should be. For they foster most strongly, in the people who should feel them least, guilt and shame. Worse, the statements of "our" universal responsibility and what "we" have done are inaccurate (despite being made by apparently intelligent people). They contain a logical fallacy that, again, is terrible in its consequences.

    I have struggled with my own feelings of shame and guilt around what is happening to Earth and the kindred life forms with which I share this planet. Most of us do, one way or another. Simply caring deeply for Earth and our non-human kin and recognizing how human action has damaged both leads inevitably to feelings of guilt . . . guilt by association that the species of which I am a part did this, guilt and shame because of my inevitable collusion in the process (for I, too, use plastics, drive a car), shame that I am a member of a species that has done such terrible things, shame for merely existing.

Millions of people feel this kind of shame and guilt. Many of them do their best to find absolution, to make things right so they can be forgiven. Some become vegetarians or practice intensive recycling, perhaps reduce their dependency on fossil fuels or buy an electric car. They give up single-use plastic bags, maybe exhort everyone around them to do the same, or become extremely sensitive to the plight of minorities or displaced tribal groups, and so on, endlessly. They do everything they can to reduce their footprint or find some kind of behavior through which to become moral – moral enough to be redeemed, to override the niggling sense of shame and guilt that will not let them go. Nevertheless, I have never seen a single one escape the guilt and shame they feel by doing so (though they may become self-righteous as time goes by). The more they try to obtain absolution, the further away from them it seems to get. (And quite often, as it does, they ramp up their behavior, try to force everyone around them to do something, too, become single-minded, single-focused, lose their sense of humor, their tenderness, their ability to love. I know . . . for once upon a time, that was me.)

This is what happens when a problem is misidentified. But the feelings of guilt and shame for what is happening to Earth, like those of grief, are trying to tell us something; they conceal crucial, important messages within them. Understanding those messages and acting on what they tell us is fundamental to the needed shifts in states of mind and being that we face as a species. But getting to those messages necessitates understanding the logical fallacies inside the words "we" and "our" – and, ultimately, doing something different than merely accepting that kind of collective responsibility.

There are three groups of people that are blended into the words "we" when people talk about what "we" have done to Earth. But not all people belong in that universal "we." This is the logical fallacy at the heart of it. This universal "we" puts responsibility where it does not belong and obscures the places it does belong. (It is a writing technique called "sleight of mouth." As with sleight of hand, it is hard to notice the trick of it.)

When any of us unthinkingly acquiesce to being included in this use of the universal "we" each of us who does so colludes in that process and the real sources of responsibility remain obscured. Those actually responsible for the damage to our world are misidentified, and because of this, individual solutions to the guilt and shame people feel never quite work. As Hanna Arendt so clearly put it, "When all are guilty, no one is; confessions of collective guilt are the best possible safeguard against the discovery of culprits, and the very magnitude of the crime the best excuse for doing nothing." (This is why, I think, the proper response to the use of universal terms in this kind of situation is a very strong and passionate "fuck you." Remaining silent risks colluding with the moral and emotional cruelty that is at the heart of assertions of collective guilt.)

Oddly enough, there is more than mere laziness underneath the careless use of such universals. Rithika Ramamurthy in *The Drift* makes an interesting and unsettling point . . .

> *In March of 2017, the American Psychological Association officially announced that feelings are valid: climate anxiety is real. . . . The preface to the report explains that beyond stress and depression, the psychic responses to climate change include fatalism, fear, helplessness, and resignation." To combat these symptoms, the authors prescribe individual solutions such as practicing mindfulness and proactive preparation. Having identified the deadly and deleterious effects of climate change (and merely thinking about them) as sources of severe mental distress, the leading experts of psychological health in this country recommend "resilience": toughening up, fostering optimism, cultivating self-regulation, and other strategies to keep your cool.*
>
> *The focus on self-care as a solution to climate anxiety tracks with the decades-long effort by global oil conglomerates to*

make climate change your problem. The "carbon footprint," an insidious invention of BP's PR consultants in 2004, centers individual consumptive practices in the fight against climate change: recycling plastic, driving electric, and going vegan are voluntary choices that contribute to lowered emissions. Investing every single personal act with planetary consequences chimes with self-care strategies for controlling and changing individual behavior – the very approaches the APA pitches as climate anxiety solutions.

The shifting of responsibility to each individual on our planet, making them responsible for what is happening, is quite an elegant use of the sleight-of-mouth technique. It obscures, as Arendt knew quite well, who is actually responsible. So, let's take a deeper look, lift the veil, and find out who is behind the curtain.

Again, there are three groups of people submerged into that universal "We." The first is:

- **Scientists, researchers, technologists, engineers, inventors.**

It breaks tremendously strong cultural injunctions to say this but it is now time, past time really, for us to look more closely at what is true and begin to accept its implications. Scientists are often well-meaning people but few of them seem to understand that the way they do their work, the paradigm through which they think, is root to the problems we face. (As well, it is important to understand that a monotheism antagonistic to an animist Earth also has a role in this. Though not well understood, reductive, mechanicalistic science is perhaps the most successful and powerful of all the protestant sects. It is anti-animist to its core.) This group, which often includes intellectuals of diverse interests and orientations, creates the theoretical structure, the epistemology, that allows, stimulates, dissection of the world, while simultaneously disseminating the widespread belief that nothing out there (except us) is intelligent (or even, in some cases, alive) . . . and that only the unsophisticated and foolish would think otherwise. Its legitimacy as the default orientation or climate of mind of a functional, modern culture is never allowed to be seriously questioned. Foundational to its approach is an allegiance to and promulgation of dissociated mentation and human abandonment of the feeling sense.

Vaclav Havel is succinct about this group's crucial involvement in the problems we face. His speech "Politics and Conscience," in fact, specifically addresses it . . .

> *To me, personally, the smokestack soiling the heavens is not just a regrettable lapse of a technology that failed to include "the ecological factor" in its calculation, one which can be easily corrected with the appropriate filter. To me it is more, the symbol of an age which seeks to transcend the boundaries of the natural world and its norms and to make it into a merely private concern, a matter of subjective preference and private feeling, of the illusions, prejudices, and whims of a "mere" individual. It is a symbol of an epoch which denies the binding importance of personal experience – including the experience of mystery and of the absolute – and displaces the personally experienced absolute as the measure of the world with a new, man-made absolute, devoid of mystery, free of the "whims" of subjectivity and, as such, impersonal and inhuman. It is the absolute of so-called objectivity: the objective, rational cognition of the scientific model of the world. Modern science, constructing its universally valid image of the world, thus crashed through the bounds of the natural world, which it can understand only as a prison of prejudices from which we must break out into the light of objectively verified truth. The natural world appears to it as no more than an unfortunate leftover from our backward ancestors, a fantasy of their childish immaturity.*

This orientation has allowed the autopsy of the natural world to proceed in earnest. Information gathered from its dissection of living beings, of Earth itself, is used by researchers, technologists, engineers, and inventors to create substances (such as plastics or agri-chemicals), machines (giant earthmoving tractors, computers, automobiles), and medical interventions (pharmaceuticals and surgical procedures) that have torn Earth and its ecosystems asunder and allowed the world's population to expand beyond the ability of the planet to support it.

I did not do these things and, almost certainly, neither did you. It is important to understand that for centuries millions of people have fought against what this first group of people have done, have questioned their underlying belief systems, have insisted on a different path. There are millions of people who have not colluded in what is happening to our world but instead fought against it with every breath they took. There are millions of us that fight still.

What is true is that without this first group of people (and without its anti-animist, monotheist underpinnings), the second group I speak of could not do what they do. And without the second group, this first group would never have prevailed in their worldview or behavior. The second group is:

- **Large (and often unrestrained) capitalist corporations and businesses (which, for me, includes the majority of the extremely rich), the legal profession, legislators.**

Large multinational corporations are, for the most part, extractive entities with long life spans and without integral morality. They utilize what the first group develops, often employing many of that group in the process, to facilitate their extractive work and increase their profits. They have rarely been and are not at this point in time ecologically oriented entities. To be so would put land ethics before profits, which few of them are or have ever been willing to do. Corporations such as these, through the power they have, buy legislators wholesale and hire the legal professionals who draw up legislation that, when passed, allow them to do what they do. They vigorously defend their actions in courts that are run by judges whom they sometimes have also bought or who have come out of their "think tanks" or political organizations. Police and military forces around the world implement the legislators' laws and the courts' decisions, in essence carrying out the will of extractive corporations and the rich irrespective of the desires of the larger populace or of ecologically sound, sustainable behavior. *I am not one of the rich, not a member of the* one percent. I do not, and have never, run one of those corporations. I am not a scientist or technologist, legislator or judge who has been bought by one. And most likely neither are you.

These first two groups are the ones actually responsible for the ecological devastation we see and feel around us every day of our lives. Hundreds of millions of people, even billions, would prefer to buy food and clothing and homes that are ecologically sustainable. They would prefer to go by train or public transport or walk to the places they need to go. (Some would even prefer to live more simply, as their ancestors did for so very long.) The majority of people would prefer to be proud of what our species does for and with the land and other living beings around them. But no individual person has the kind of power necessary to stop these first two groups from doing what they do. Blending everyone into the universal "we" creates, as it is meant to do, a sense of disempowerment as well as a feeling of personal responsibility for the damage in the very people who are least responsible for what is happening. It shifts the responsibility from those who are actually responsible to those who are not ("Be responsible: Stop using single-use plastic bags!").

The understanding that we, as individuals, are in a different category is essential: it is the way out of our disempowerment. *We* are part of a different "we."

We are . . .

- **We the people.**

Those of us in this group are born into a human scenario that we did not create and to which we, as children, have had to adapt as all people have had to do since people have been. We deal with what has been set before us and it is only as we age, as we move out of childhood, that we begin to look around ourselves and notice that there might be some problems with the human scenario in which we are living. Some of us then begin the hard work of trying to change it – and the even harder work of changing ourselves. Nearly all of us would like it to change, for it to become more humane, for those first two groups to care more for Earth and future generations. Nevertheless, we are still faced with the problem of living – every day of our lives. We hunger and need food, we need and must find shelter, and like all people who have ever been we yearn for love and family and companionship. We adapt to the system . . . or we die. And so, we do the best we can.

It is this group of people who should never be included in the "we" that has created the anthropocene – doing so conceals the actual sources of our troubles. Responsibility is put in a place it does not belong. And the media, academics, writers, environmentalists, and activists should shut the fuck up about the universal "we" and put responsibility where it actually belongs. When blame is placed correctly, it clarifies what has to be done; it focuses effective action and reveals what is true rather than obscuring it. If those first two groups were forced to significantly alter their behavior, the problems that Earth and our species face would dwindle considerably. They might even be manageable. Those groups are, however, quite clever in diverting attention away from themselves; the trouble is, as it historically often is, that so many people keep falling for it. As Hannah Arendt so clearly saw: "confessions of collective guilt are the best possible safeguard against the discovery of culprits."

There are, of course, things that each of us have done for which we feel shame and guilt. But it is *only* for those that we are responsible. (Like Elie Wiesel, I do not believe in collective guilt. It is in fact a fundamentally evil concept. It is closely interwoven with the concept of blood purity and its reverse, blood contamination, that is, a dangerous trait or character defect held to be integral to a race or species. It is a concept just as evil as collective guilt, if not more so.)

It is often a very hard thing to understand and accept that we are only responsible for the things we have actually done (and not for what our parents or our race or our culture or our gender or those first two groups have done). Accepting responsibility for our own acts and then doing something about them is within the reach of each and every one of us. *These are the things that each one of us can actually do something about.* Some of them really do connect to the damage that has been done to Earth and our kindred species (just as others do not).

I will tell you one of mine . . .

When I was twenty, in 1972, I moved to the high mountains of Colorado. The "back to the land" movement was strong and it called me just as it was calling thousands of others. For most of my life I had felt an urgency to live on wild land and finally, in that year, I could no longer ignore it. I'd heard a rumor that there were abandoned homesteads in the mountains and that if the owners were found and asked, they would sometimes agree to a person repairing and living in them for free. And so, this is what I did.

The place I found was a seven-by-fourteen-foot cabin, long abandoned. But it was still sound enough that with a little work it could be made habitable. And so I set about repairing it. As I worked, I noticed that there was a small hole on both the inside and outside of one of the walls. Looking within, I found that a small bird had built her nest there. Oddly enough, she wasn't worried about me, in fact she was deeply agreeable to my presence and we soon became friends.

The head of my bed was near that opening and as I lay there in the evenings reading (as I do and have done each and every evening of my reading life), she would come partway out and turn her head to the side, looking at me with those bright eyes of hers. She would talk to me sometimes, in a language that only she and her kind understood, about her day and how things were going for her. Then, in the language I understood, I talked back to her, responding to her stories, telling her of my day and life. And for some reason both of us came, over time, to understand each other and our friendship deepened.

She wasn't the only nonhuman life that had found a home in that cabin, of course. As time went by I discovered that mice were living there too. They kept getting into my food and leaving little pellets of poop everywhere and finally I realized I needed to do something about it. So, one day I borrowed the neighbor's cat and brought him home with me. I was young then, so very young, and often I did not think things through.

Late that first night, while I was deep asleep, I heard that tiny bird scream. And I knew as I startled awake what had happened. I leaped up, grabbed the cat, and threw it out the door. Then I found that tiny body, broken and bloody on the floor. And I knelt there, weeping.

As I write these words, and despite all that I know of my youthful ignorance and lack of ill intent, that pain is still as fresh and sharp in me as the day she was killed. It comes to visit me from time to time, often in the depths of night, when the darkness awakens me and says, "There's something we need to talk about." I so wish I could go back and stop that young man from bringing that cat into that cabin. For I still excruciatingly feel the death of that tiny bird who welcomed me into her life, who trusted me enough to become friends, and who had shared so many stories of her days with me. I feel guilt for the act, shame for my existence, for it is that existence that brought to that bird her death.

This is what personal shame and guilt feel like, what they *are* like. And a great many of us feel these kinds of things for something we have done. We feel them every day of our lives. (And, regrettably, the numbers of events like these accumulate through the years of our lives. They become more poignant, and painful, as they do so – for the moral sensitivity of the heart grows with every year of our lives. In elder years, they cut like the sharpest knife you have ever known.)

Importantly, there is something that decades later I found to be true about that moment in my life. I think perhaps it is something that each and every one of us is meant to find as we journey through the years. It took me so long to find it because, of course, as I often do, I looked in all the wrong places. I kept replaying the memory, feeling the pain of it, castigating myself for my ignorance, for being the agent of that tiny bird's death. Then, in pain and grief, I actively pushed the event away from me, tried to turn back time to the moment when I was still innocent, to the time I had not been the agent of another being's death. (Far too many of us, I have learned, equate a state of innocence with being moral.)

It has never worked, it will never work. That past state of innocence can never be regained. When I finally understood that, I did something different. I stopped pushing it away. The moment I did so my experience of that moment changed. My orientation shifted, I started to see in a different way – from that tiny bird's point of view. At the moment of her death that tiny bird, odd as this might sound, came to live inside me – and every day afterward she'd been trying to tell me something important. That is why I could never get rid of her, why her death kept returning to me in the night. So I stopped resisting her presence and welcomed her into my depths, into the fabric of me. She took her place then, becoming one of the inner voices that I carry inside.

That little bird is always with me now, sitting on my right shoulder, looking out at the world with her bright, inquisitive eyes. When I am about to make a choice, to decide on some particular behavior that I perhaps might not have thought through well enough, she turns to my ear and speaks insistently in her beautiful voice. I stop what I am doing and – for a time – her seeing becomes my seeing. A different point of view replaces my own. It is then I am able to perceive what might come from the choice I'm about to make, possible outcomes that single-vision has kept invisible to me.

That sudden moment of in-sight revealed that my shame and guilt (and my continual attempts to regain my lost innocence to avoid those feelings) had concealed a companionship I needed in order to become more fully human. I was so caught up in feeling bad that I couldn't see what was being asked of me, that it was important for me to allow the dead to enter inside and take their rightful place.

*It was important for me
to allow the dead to enter inside
and take their rightful place.*

What a strange idea that is. It conflicts with so many things most of us have been taught about life and this world. About death and our relation to the dead. I suppose that's why it took so long for me to understand. And perhaps as well there was some sort of lingering sense that if I did let the dead in, I would forever be judged by them once I had.

Oddly enough, my experience has been quite the opposite.

The dead that come to me in the night (and there are far more of them than that tiny bird) have been trying through the years to tell me a great many things that I've needed to know. A crucial one is that the moral sense within each one of us develops over time. It is only through our errors, *and the feelings that the impact of those errors on others engender in us*, that our moral self matures. The dead, who are more with us than our western cultures admit, know that it is part of the function of elders, alive *and* dead, to carry this developed moral sense for our cultures, for the young, even, quite often, for the middle-aged who are often not so mature as is believed. For the young are still learning the difficult lessons of the arrogance and psychological myopia with which all of us are born. Understanding that the moral sense only develops over time, that it can only develop through errors of judgment and the pain those errors inflict on other lives – *and that this is true for every person there is* – allowed the beginnings of self-forgiveness for me.

Each of us, I am pretty sure, can only know ourselves in context, in relation to the scenario in which we live, in relation to the people among whom we move, and never in isolation.

The acts I have done are not forgotten, the pain is not lost. But they are now integrated into the complexity of who I have become over the decades of my life. They are part of who and what I am. Becoming whole necessitates the integration of both the good and the bad (and everything in between) that has been done in a lifetime. (*Integrity* is the state of being whole and undivided.) All that, along with habits of mind, tendencies of character, aspects of soul, and so much more, when blended together as a unique whole are what make up the fabric of the self. It is not possible to keep this and get rid of that, to keep only the good and get rid of the bad. To try to do so only deforms the self and causes the people around us unrelenting harm. (That is, after all, the teaching of Robert Louis Stevenson's *Dr. Jekyll and Mr. Hyde* and Oscar Wilde's *The Picture of Dorian Grey*.) All that can be done is to go on with more awareness than before while carrying the pain as well as the gifts and teachings that all those errors in judgment have brought – but fully integrated now, integral, unremovable from the fabric of the self.

And so I found the absolution I sought – though of course the absolution I found bears little resemblance to what I had imagined absolution to be. It is not a return to a state of innocence – as so many people erroneously believe, or wish, it will be – but an integrated expansion of the self outward into the livingness of the world – and that always includes pain . . . and all the memories, and the errors in judgment, that will never go away. It is an expansion into truths that Earth brings us every day of our lives, whether we wish to know them or not. And with that expansion comes a maturity of responsibility that I think can be found no other way.

The death of that bird was caused by something I personally did and the shame and guilt belong to me. I truly was responsible – whether I intended what happened or not. And because I was, the actions needed to make it right were mine alone. Generalized feelings of shame and guilt, the ones stimulated by a sense of collective responsibility for the ecological damage to our home, are something else. Few of us are personally responsible for that ecological damage. I am not responsible for that ecological damage. You are not responsible for that ecological damage.

Yet, just as it was with the feelings I had about the death of that tiny bird, there is some truth (or truths perhaps) hiding inside those larger feelings of *collective* shame and guilt. There is a reason they keep returning to the heart over and over again and disturbing the equanimity of the self. (Such feelings are very easy to leverage by outside forces who do not wish us well. It behooves each of us to make it difficult for them to do so.)

Again, as with that tiny bird, there came a day for me when I decided to quit pushing the feelings away. I turned toward them instead and let them become fully present inside me. I began to sit with them, contemplate them, feeling more deeply into their depths. I immersed myself in the feelings but did not castigate myself, did nothing except let the feelings have their life inside me while seeking the truths, and teachings, they were trying so hard for me to understand.

Over time, as the years progressed, I did, finally, understand. And I found that there are two initial truths/messages/lessons/teachings inside those generalized feelings of shame and guilt. (There are two others that are inherent in these first two but they took me much longer to find and I will talk about them in a little while.) Interestingly, the truths I found don't have anything to do with collective responsibility. They are concerned with something else entirely.

The first of those truths is this . . . *because we are alive other life will die. Because I am alive other life must and will die.*

We are, all of us, killers – inadvertent, intentional, and by association – and *nothing will ever change that fact*. Because *you* are alive, other life will die. That is a foundational ecological reality; it's an integral aspect of the scenario from which we have been expressed and in which we are embedded. There is nothing to be done about it. Nor is there anything in it to be ashamed of. That we are killers is true of all human beings. It has always been true. It will always be true.

> *Think about it ... it's an important question:*
> *What are you so afraid of?*
> *What will happen*
> *if you accept that you are a killer?*
> *What will happen if the people around you*
> *find out that you are one?*

Because so many of us so fervently deny this aspect of our nature it becomes impossible to integrate and take responsibility for the fact of it. From that denial, terrible shadows emerge into the world, darknesses we do not allow ourselves to see.

It was the stories of those who are known as "accidental killers" that finally broke through my unwillingness to understand. And as always with the answers I seek, I came upon this one by chance one day when I thought I was just living any old day at all.

There are a great many more "accidental killers" than people suppose. Most of us have seen them mentioned in media articles at one time or another. Their stories are factually different but in essence pretty much the same. Here is one of them . . .

> *Once Upon A Time, there was a young woman who was driving home to get something she forgot and that she wanted to take to her new apartment.*
>
> *She is driving down a street with cars parked closely on each side when suddenly a small child runs in front of her in just such a way that she hits him before she even registers he is there. She enters a strange state of shock; time suspends itself. The world*

around her becomes dreamlike, unreal, as if this can't really be happening. Then, inevitably, she feels the double bump of her tires going over that tiny body. She stops just as the boy's young mother runs into the street. Stepping outside the car, she watches the woman hold her shattered child in her arms, watches as she begins to scream the loss that is now running through her heart, her life. That will forever run through her heart and life. It's a scream the young woman will hear, over and over again, every day of her life from that moment on.

Later, the young woman's friends tell her it was not her fault (which it wasn't). Yet she knows a truth they cannot bring themselves to accept: she is, and always will be, the agent of the boy's death – if she had not been driving down that street at that moment in time, the boy would still be alive. Nothing will ever erase that truth.

Her old life (that unwoke, unaware, innocent life) ended the day she killed the boy and she has not yet found a new one. Her struggles, her depressions, her self-castigation, the endless questions, the unremitting thoughts: Was the thing she'd forgotten to take with her worth a child's life? Why had she taken that road and not another? If only she had taken another route, the boy would still be alive. If only she had been paying closer attention, maybe she could have stopped in time.

Her nightmares (despite all the therapy) have gone on for years. Every day she wakes unable to escape the fact that she is a killer of small children – whether she meant to be or not.

The real world is telling her a truth that most everyone in america is doing their best to hide from. Each and every one of us has run over little boys but we usually do it in ways subtle, or apparently different, enough for us to ignore. So we continue to believe untruths about ourselves (and make the disney corporation billions in the process).

Eventually, the young woman began to write (and publish on her blog) about what had happened to her. Scores, perhaps hundreds, of accidental killers soon read what she had written.

> *They had found, at last, someone who was talking about what had happened to them also, someone struggling to bring this difficult truth out of the darkness in which our culture has hidden it, into the real world, into being, into the light. Someone who was struggling to give it form so that this thing that all of us are hiding from can somehow become integrated into our cultural life and into ourselves. Slowly, stumbling over their words, they too began to write, adding their own experiences to the story that brave young woman had had the courage to speak out loud.*
>
> *Maybe, I sometimes think, if enough of us speak out and tell our own stories about the moments we killed a young child (or the tiny, happy bird who trusted us with its life) then maybe our country won't have to go overseas and kill children who don't look or talk like us. Maybe we won't continue to destroy the lives and habitats of millions of our kindred species who have done us no harm.*
>
> *Still . . . I wake in the night sometimes and the young woman's story comes alive inside me, the scene opens itself and plays out on the screen of my inner vision. I always try (as I am sure that young woman has tried) to stop the boy, to stop the car, to stop the past from becoming real. But the bumps still come. And in the darkness of night I still see the mother screaming and holding her broken son.*
>
> *Earth, I have realized, will never give up trying to get me to understand and integrate these truths. It loves me, far too much to allow me to remain in the dark.*

And, now, I am sure someone somewhere is saying: Am I, Stephen, implying that that tiny bird and that boy are the same? It's the wrong question. The more appropriate one is this: Ecologically, to Earth, are the living beings of this planet organized by a hierarchy of value, that is, are human beings more important than all other life forms that are or have ever been? The answer is, quite obviously and simply: *No, they are not.* Every species, every individual, that is expressed out of the ecological matrix of the planet is done so for a specific ecological purpose. None is more important than any other, none higher, none lower. (Darwin was insistent about this – though most of his followers conveniently forget it. He had a sign in his office to remind him: *There is no higher or lower.*) There is no evolutionary escalator or pyramid with us at the top, the crowning form of life in all of creation. That human beings believe, that you yourself might believe, probably do believe, that we are is part of an exceptionalist belief system that is itself a root cause of the ecological problems our species faces.

> *And now perhaps you sense the edges of one of the most important questions that human beings have refused to ask themselves for far too long: What is the ecological function of the human species?*
>
> *And if the answer that occurs to you now has anything to do with the human species being special, something as foolish perhaps as "we are the first time the universe has become conscious of itself," or maybe "we are the only intelligent species on this planet," or even "we are the only species that possesses a soul," it's wrong. We are the same as everything else, each expressed for an ecological purpose. We are no more – and no less – important than an ant, a bacterium, a redwood, a tiny bird, or the mosquito that you just killed. Ecologically, there is **no** higher or lower.*

Each and every one of us is a killer; *Earth needs us to understand this and come to terms with it*. It's an essential truth of this world, of the ecology of Earth, of the scenario that has expressed us into being. And that truth is foundational to our successful habitation of Earth, for only when we accept it can we learn to be responsible for it. Hidden, it lives a life of its own. Hidden, we remain a good doctor, concerned only for the welfare of others, but another aspect of ourselves, in the darkness of night, travels elsewhere in the city and the next day when we awaken there is news of another death.

That each and every one of us is a killer is a hard thing to accept and integrate into self-identity. It conflicts with the either/or belief system of good and evil that most of us have internalized. Nevertheless, life is far more complex and nuanced than that simplistic, monotheistic view would have it. Moving through the world while being oblivious to the killings we commit (because we are good people, we are not killers) interferes with the development of our moral selves, of the responsibility we have for the death of other lives. Out of that responsibility, once it is accepted, comes a different kind of behavior. One that is essential to sustainable habitation of Earth. One that is essential to sustainable habitation of ourselves.

Now another memory begins to play itself out on the screen of my inner vision. I see the opening scenes of the movie *Last of the Mohicans*. Three men are running through deep and old forest. One of them, Hawkeye (Daniel Day-Lewis), stops and fires a musket at the elk they have been hunting/chasing through the forest. It staggers, then falls into a ravine. The three men, Hawkeye, Uncas (Eric Schweig), and Chingachgook (Russell Means), walk to where it has fallen. They stop, clearly in reverence, then Chingachgook says in his native language, "We are sorry to kill you, brother. We do honor to your courage, speed, your strength." He does not say, but it is implied that, "We kill you so that we might live."

And then as so often happens, this golden thread of memory leads to another that arises of its own accord. It arises out of memories thirty years old. It's as clear in my experience, in my mind's eye, in the feelings of my heart, as the day it happened.

I am driving my son to school; he is very young, perhaps six or seven. We are traveling down Flagstaff Mountain Road from our home nine thousand feet above the sea into Boulder, colorado. The land where we live is as it has been for the past thirty thousand years. It has not been farmed nor ranched nor logged. It is *alive* and wild and filled with presences that most americans insist do not exist.

We are traveling through a dense and wooded forest, the road curving between the trees, when I glimpse from the corner of my eye a herd of perhaps forty deer running through the forest. They are as silent as drifting fog, their coats the color of late fall grass. They blend into, are part of, the landscape they move through, as if they are landscape transformed.

One of them, a young adolescent male, intersects the front of our car with perfect timing. There is no way to stop and so we hit him. I hear/feel the multiple thuds/bumps as he's rolled over and over underneath the car. With each I feel the car lift as the wheels move up and over his body. All of it happens very fast. I stop the car. I'm shaking.

My son and I get out and walk back to where the young buck lies on the ground. He's still alive but terribly wounded. We kneel beside him, one on either side of his head. His eyes are wide, the whites very visible. He's terrified. I see/feel in him a life not so different than my own or any young man I have known. Around me I sense, see by the unworn sides of my eyes, the deer scattered among the trees. They have stopped, as softly present now as thistledown held motionless in air, watching, waiting, unmoving. Some strange kind of expectancy has come over the land, over all of us, almost as if the world itself were holding its breath. I have never felt anything like it before.

My son and I lay our hands along the sides of the deer's head. I begin to speak softly, "It's alright, you are not alone, I am sorry to have done this to you."

The young deer breathes raggedly, blood running out of his nose. His eyes widen further, the whites become more pronounced. His face fills with fear of what's approaching. We are, all of us, caught in suspended time, my son and I, the dying deer, his companions – the herd – held motionless in the kind of moment that only happens when a great bell has finished its ringing and sound has become a silence penetrating everything, filling the world with a meaning that can occur no other way.

Then, slowly, the tension begins to leave the young deer's body, his eyes dim, and with a last ragged shudder his breath staggers, then stops. I feel something leave his body then, its weight as intangible as sunlight on summer skin but as real as the presence of my beloved beside me. It passes through the forest like a sigh, an almost imperceptible wind. And everything that is feels it and knows its meaning. Then, a tiny instant more, and it's gone.

And I *know* in that moment that the herd of deer understand what that nearly imperceptible breath that has moved through the world is and what its passing means – but they have a much different relationship with it than I do. They stand for a time, silent, engaged witnesses. Then, as one, without any observable signal, the exact moment it passes, they move again, and are gone.

Time once again begins its forward flow. My son and I pick up the young deer and put him in the back of the car, turn and drive him home.

My wife takes him from us and then we turn and drive into town to meetings that cannot not be postponed. While we are gone, she unmakes him, prepares him as food that will feed us in winter. When she is done, she places what remained on the edge of the forest in an aspen grove we can see from the house. The ravens and magpies come first, then the eagles, driving the magpies and ravens away, then the turkey buzzards, driving the eagles away, then the coyotes, taking what is left for their own. When they are done all that is left is the long tubular-shaped remains of the grasses that the young deer had eaten but not yet digested. Even the intestinal membrane surrounding the grasses has disappeared. Before long the grasses, too, are gone, taken back into Earth, and nothing remains but the memory and the life his body gave us in the food we eat as the winter months pass by.

All of us must kill in order to live – and all of us inadvertently kill a great many things that we do not eat as we travel through life. When we know that, *when we truly understand* that other, *sentient, intelligent* life must die, will die, does die simply because we exist (and this most definitely includes every plant that we eat or kill), a certain kind of humility comes into being that can be found no other way. And that humility engenders, as it is supposed to do, an acute sensitivity to, even reverence for, the lives that are taken – whether intentionally or inadvertently or by our association with those who kill for us every day of our lives – whether they are killing the food we eat or the trees that make our houses.

It is only when we understand these things and accept them into the fabric of who we are as foundational, inescapable truth that it becomes possible to sincerely kneel and pray, to tell our kin of our sorrow for their death, to thank them for giving their life so that our own may continue or to apologize to them for our being the inadvertent or associational agent of their death.

Simply because we are alive, other life will die.
This is a fundamental truth of the scenario in which we are embedded.
It is an ecological reality that can never be escaped.
And there are reasons for it being this way,
reasons of long evolutionary standing.
There is no stain upon us because of it.
That most of us in the west do not know this,
that this foundational truth is not integrated into our lives,
reveals just how dissociated from the real world we have become.
It is a symptom of the pathology of our times.
We have taken inside us inaccurate beliefs about ourselves and the world.
This is a reality we can do something about – if we are willing to.
It is a crucial step in the decolonization of our souls, our minds, our hearts.

This is one of the essential understandings that has been concealed inside the guilt and shame that plagues us. It is part of what those feelings are trying to tell us. (To understand and integrate this necessitates movement out of the simplistic moralities and sophomoric philosophies of being that we have internalized as true about our lives and world.) Understanding and integrating this is integral to the absolution we seek for it embeds us within what is real, within what we truly are, within this ecological scenario that we call Earth. And inside the reality of what we are, inside that *truth*, are the beginnings of self-forgiveness. For self-forgiveness is rooted in compassion for who and what we are, for who and what everyone that is and ever has been, *is*. There is no inherent guilt that accrues to the fact that other life must die for us to live. None.

> *This does not, however, include or excuse those groups of humans, those corporations, those professions that engage in wanton destruction of the living beings of this planet for personal gain, of whatever sort – academic, or financial, or simply from the love of killing. To them, guilt, and responsibility, does accrue. As Barry Lopez once put it, "Men smelling of evil have inquired into the purchase of our homes." Such people are never morally inert.*

It's an alien thing for us in the west at this time in our history to understand this, to feel in this way, and subsequently, to pray in this way. For we (and I am most definitely a part of this particular, universal we) have been taught that there is no soul or intelligence or companionship out there, that we are the only intelligent species on this planet. We have been taught that we can take any part of the insentient world around us as our own, as our right, and for it we need give no prayer or respect or reverence. We have been taught that we have an inherent right to engage in the exploitation of matter. (Nevertheless, some deeper, wiser part of many of us realizes there is something wrong in this, that there is another truth about the livingness of the world than the one we have accepted – this is, to a great extent, where the lingering, inescapable feelings of guilt and shame come from and what they are trying to tell us.) To believe and behave differently goes against all that reductive scientists believe, all that **monotheists insist is true, all that our culture tells us, each and every day** of our lives. But these are some of the most important truths that Earth is trying to teach us.

We fail to understand them at our peril.

I have said there are two initial truths/lessons/messages inside these generalized feelings of shame and guilt. The second one is far more difficult to grasp, certainly more difficult to understand and integrate. It undermines our ingrained sense of uniqueness, our exceptionalism, our exemptionalism in the most direct way possible. The person who has spoken most profoundly of this is the australian environmentalist and writer Val Plumwood.

As the story begins, Plumwood is in the wilderness of the northern territory of australia. She has decided to borrow a canoe from the park service and travel among the small rivers and lagoons of kakadu national park. As she is readying the canoe, the ranger strongly suggests that she stay in the backwaters, telling her, "Don't go onto the main river channel. The current's too swift, and if you get into trouble there are crocodiles."

As Plumwood comments, that first day I "glutted myself on the magical beauty and bird life of the lily lagoons." She'd had such a good time that she decides to deepen the experience the following day. She's looking for some aboriginal rock art that she knows is up one of the side channels she will pass. But as the day progresses it begins to drizzle, then to rain. The world around her is no longer magical, it's simply dreary and difficult. To make things worse, she soon realizes she is lost.

> *After hours of searching the maze of shallow channels in the swamp, I had not found the clear channel leading to the rock art site, as shown on the ranger's sketch map. When I pulled my canoe over in driving rain to a rock outcrop for a hasty, sodden lunch, I experienced the unfamiliar sensation of being watched. Having never been one for timidity, in philosophy or in life, I decided, rather than return defeated to my sticky trailer, to explore a deep clear channel closer to the river I had traveled along the previous day.*

And so she, as many of us do, fails to listen to what her deep, feeling self is telling her.

The rain falls harder, the wind comes up. Her canoe rides only a few inches above the surface of the water; it begins to fill. Every so often, she pulls over and tips water from the canoe. The day grows increasingly miserable – and so does she.

Despite the map and her normally reliable sense of direction, she remains lost. So, finally, she pulls to the bank to try and get her bearings. As she walks up and onto the bank "the feeling of unease that had been with me all day intensified." But once again she dismisses the feeling and continues on. She walks farther, crests a dune, and is shocked to find she is only a hundred yards from the main river channel. It would be quite easy to use it to travel back to her trailer and by this time she's so miserable that, despite the ranger's warnings, she pulls the canoe into the river and sets out.

Five or ten minutes later she notices that what she had first taken to be a floating log has developed eyes. She turns the canoe away from what she now knows is a crocodile. But the crocodile has other plans.

> *Although I was paddling to miss the crocodile, our paths were strangely convergent. I knew it would be close, but I was totally unprepared for the great blow when it struck the canoe, again it struck, again and again, now from behind, shuddering the flimsy craft.*

Realizing that it will only be a short time before the crocodile capsizes the canoe she decides to get to the bank and leap to safety if she can.

> *The only obvious avenue of escape was a paperbark tree near the muddy bank wall. . . . I steered to the tree and stood up to jump. At the same instant the crocodile rushed up alongside the canoe, and its beautiful, flecked golden eyes looked straight into mine. . . . I tensed for the jump and leapt. Before my foot even tripped the first branch, I had a blurred, incredulous vision of great toothed jaws bursting from the water. Then I was seized between the legs in a red-hot pincer grip and whirled into the suffocating wet darkness.*

It is here that the great insight Plumwood brings to the world begins to emerge.

> *Our final thoughts during near-death experiences can tell us much about our frameworks of subjectivity. A framework capable of sustaining action and purpose must, I think, view the world "from the inside," structured to sustain the concept of a continuing, narrative self; we remake the world in that way as our own, investing it with meaning, reconceiving it as sane, survivable, amendable to hope and resolution. The lack of fit between this subject-centered version and reality comes into play in extreme moments. In its final, frantic attempts to protect itself from the knowledge that threatens the narrative framework, the mind can instantaneously fabricate terminal doubt of extravagant proportions: This is not really happening. This is a nightmare from which I will soon awake. This*

> *desperate delusion split apart as I hit the water. In that flash, I glimpsed the world for the first time "from the outside," as a world no longer my own, an unrecognizable bleak landscape composed of raw necessity, indifferent to my life or death.*

The crocodile pulls her under and begins what is known as the death roll . . .

> *Few of those who have experienced the crocodile's death roll have lived to describe it. It is, essentially, an experience beyond the words of total terror. The crocodile's breathing and heart metabolism are not suited to prolonged struggle, so the roll is an intense burst of power designed to overcome the victim's resistance quickly. The crocodile then holds the feebly struggling prey underwater until it drowns. The roll was a centrifuge of boiling blackness that lasted for an eternity, beyond endurance, but when I seemed all but finished, the rolling suddenly stopped. My feet touched bottom, my head broke the surface, and, coughing, I sucked at air, amazed to be alive. The crocodile still had me in its pincer grip between the legs. I had just begun to weep for the prospects of my mangled body when the crocodile pitched me suddenly into a second death roll.*

Astonishingly, the crocodile brings her up again, and for some reason its jaws relax. Plumwood finds herself near a large branch of the sandpaper (paperbark) fig tree growing on the bank of the river. She grips it and pulls herself up, trying to get behind the tree, to once more climb to safety.

> *As in the repetition of a nightmare, the horror of my first escape attempt was repeated. As I leapt onto the same branch, the crocodile seized me again, this time around the upper left thigh, and pulled me under. Like the others, the third roll stopped, and we came up next to the sandpiper fig branch again. I was growing weaker, but I could see the crocodile taking a long time*

> to kill me this way. I prayed for a quick finish and decided to provoke it by attacking with my hands. Feeling back behind me along the head, I encountered two lumps. Thinking I had the eye sockets, I jabbed my thumbs into them with all my might. They slid into warm unresisting holes . . . and the crocodile did not so much as flinch. In despair, I grabbed the branch again. And once again, after a time, I felt the crocodile jaws relax, and I pulled free.

She escapes then, terribly wounded. Oddly enough her groin, because of the type of shorts she is wearing, is not as badly damaged as she'd feared. It is the thigh that has suffered most. "The left thigh hung open, with bits of fat, tendon, and muscle showing." She tears strips from her clothes and binds the wound, begins to try and make her way back to the ranger station. Luckily, the ranger, realizing she had not returned, had set out to find her. He hears her cries for help but cannot reach her from where he is and so returns to the ranger station for more assistance. Eventually a rescue craft follows the river, finds her, and takes her to darwin hospital for treatment that will last more than a year.

The lesson she learns, that all of us must learn, is that *we, too, are prey. And we always will be.* We are meant to be killed, to be eaten just as we are meant to kill and to eat. We are no more special or important than any other life form. This is a fundamental truth of Earth; it can never be escaped no matter how strongly our cultures, or we, try to escape it. We are not isolated intelligences, the peak of evolution, exempt from the realities that govern life here. And it is this truth Plumwood learns in the hardest way it can be learned.

> *Before the encounter, it was as if I saw the whole universe as framed by my own narrative, as though the two were joined perfectly and seamlessly together. As my own narrative and the larger story were ripped apart, I glimpsed a shockingly indifferent world in which I had no more significance than any other edible being. The thought, "This can't be happening to me, I'm a human being. I am more than just food!" was one component*

of my terminal incredulity. It was a shocking reduction, from a complex human being to a mere piece of meat.

Our failure to understand that we are the eaten as well as those who eat is the strongest indication of our lack of understanding of our embeddedness in this ecological scenario we call Earth. The foundational truth of this planet *is* ecological. It *cannot* be escaped no matter what technological, utopianistic (or monotheistic) fantasies are put forth asserting that we can. We are ecological beings on an ecological planet. We are expressed out of Earth, we live embedded within that scenario, are *of* that scenario, and to Earth all of us, sooner or later, will return. Like all life, we are both the killers and the prey. We are meant to be eaten, sooner or later, by one thing or another. Sometimes it is a crocodile, a species that has been here for some one hundred million years; other times it is something as invisible to the eye as a viral pathogen whose species is even older than that. To them we have been here but the blink of an eye.

And no, neither Earth nor Universe is killing us for our sins, individually or as a species. The idea of a vengeful god comes, in its present form, out of the old testament writings and insistences of a monotheistic christianity, islam, and Judaism, not out of the reality of Earth or Universe. Ecological truths, even ecological repercussions to such things as overpopulation or the environmental devastation from industrial resource extraction, are no different than dropping a rock on your toe. Gravity has no personal feelings in the matter. Exceeding ecological limits of necessity generates ecological adjustments. It's not personal. Nor is it personal when we are eaten. It's just the way it is, the way it's always been, the way it will always be. Death is built into the system. And it's built into the system for a reason. We can disapprove of that as much as we want; it won't change anything.

That we are meant to be eaten is an essential aspect of our biodegradability. Death comes to all life and from that death all life comes. Without death, without that integral biodegradability in each and every living being, the ecological fabric of our planet unravels. Without death there can be no new life, no new generations, no trees, no forests, no people, no-thing at all. Our dying (and the diseases we suffer on the way) is neither payment for our sins nor "ecological facism" as some writers would have it (writers who believe we deserve to and should, by our nature, escape all ecological and biological realities), but one of the root truths of life itself. We are, forever and always, ecological beings on an ecological planet; there is no escape from that reality. In the most expansive sense possible, we carry that truth with us wherever we go, to Minneapolis, to Mauritania, or to Mars.

The simple fact of our birth means that we will kill, perhaps inadvertently as I and that young woman have done, or intentionally in order to have the food we need – or merely because we desire relief from the insistent attentions of an annoying mosquito or fly. Sometimes it will be by association, through the houses we need for shelter, the clothes we wear to survive the seasons, the food we buy in supermarkets. Some part of the natural world always dies so that we might live. If we understand that and understand as well that *everything that is, is not simply meat but is, like us, more than meat,* then we, by necessity, begin to approach that killing with reverence and humility. It is this quality of humility that is necessary for our sustainable habitation of this world. And that humility also includes the understanding that we are each given only a little time here, that each of us will grow old, that we will biodegrade, will suffer disease, will die so that other life might live, so that this 4.5-billion-year-old scenario we call Earth – which is far older and wiser than our species (and our beliefs) – will continue. In no other way will our species endure.

As Plumwood said, many years later, in an article published just after she herself had died ("Tasteless: Toward a Food-Based Approach to Death")...

> *Since then it has seemed to me that our worldview denies the most basic feature of animal existence on planet earth – that we are food and through death we nourish others. The food/death perspective, so familiar to our ancestors, is something the*

> *human exceptionalism of western modernity has structured out of life.... Predation on humans is [considered] monstrous, exceptionalised and subject to extreme retaliation.... Dominant concepts of human identity position humans outside and above the food chain, not as part of the feast in a chain of reciprocity. Animals can be our food, but we can never be their food. Human exceptionalism positions us as the eaters of others who are never themselves eaten.... [But] by understanding life as in circulation, as a gift from a community of ancestors, we can see death as a recycling, a flowing on into an ecological and ancestral community of origins.... The food/death imaginary we have lost touch with is a key to re-imagining ourselves ecologically, as members of a larger earth community of radical equality, mutual nurturance and support.*

The complex wisdom encoded in these insights, if integrated, firmly entrenches us in the ecological food webs of this planet – in the *life* of this planet. We are not special and we never have been. The pronouncements of monotheists and rationalists that we are is (and always has been) one of the greatest errors our species has ever made. The widespread belief in human exceptionalism, in human exemptionalism, and the widespread human dissociation from our kindred life forms have each played a role in creating what our species now faces. And yes, this is something that each of us can individually do something about. We don't have to keep believing those untruths or remaining dissociated from our feeling self or from the world around us. We can choose to decolonize ourselves.

The solution to the pervasive feelings of shame and guilt that are so common now necessitates a difficult reworking of deeply embedded belief systems in each and every one of us. It isn't easy. But what is true is that if you can understand that you are personally not responsible for the actions/impacts of scientists, technologists, engineers, and corporate capitalists upon the living world around us, you take the first step.

Then . . . if you can learn to live with the awareness that simply because you are alive other life will die, you take the next step – and perhaps learn in the process to stop judging yourself for an ecological reality that came into being long before you yourself were born.

And finally, if you can grasp (and learn to not mind, to not be afraid of this truth) that you are meat and that despite your dreams, your memories, your hopes, you will always be meat sooner or later, a different kind of life can begin.

That different kind of life is one in which the limits within which all life lives are understood – and accepted – without resentment, without rage, without blame as also applying to us, individually and collectively. It is a way of life in which reverence, humility, self-forgiveness, and a deep thankfulness take the place of shame and guilt. (And over time, much of the fear of death that so many of us in the west live with might also begin to fade.)

In contrast to what far too many people believe, I know that Earth cares for us as Earth cares for all life it gives birth do. At the core, nature is not, as reductionists would have it, red in tooth and claw. Earth is more a place of cooperation than it is competition. But it is also true that each of us is intended to be folded back into the soil of this planet, that our death is meant to bring forth new life, as the death of all life does. It's just that Earth doesn't have the same psychological difficulties (that is, issues, baggage, fears, phobias, terrors) with death and pain and suffering that we do, that most "civilized" cultures do. Death and the pain and suffering that go with it, just like birth and love and beauty, are merely part of the ecological realities that are inescapable here. They *are*. And they always will be. What we do have is the choice of how we approach them and how we respond to them. That is something each of us will always have. We have the choice of what kind of person we want to be.

The question is . . . are we willing to know it?

INTERLUDE THREE

"People Possess Four Things That Are No Good at Sea"

While it is true that stones are hard they cannot survive the patience of water.

Taoist proverb

How many nights now has the stream told you, "This is the way to deal with obstacles."

Dale Pendell

I remember the first time I heard of the poet Robert Bly. It was the year I'd finally accepted the fact that I did not fit into the normal curricula at american colleges and universities and decided to do something different. I'd already taken what people call a gap year, though for me it lasted five and I called it "time off to think about stuff." During those years I'd heard about the university-without-walls system and it just so happened that there was one in the town where I lived and so, in the fall of 1979, when I was twenty-seven years old, I enrolled.

I remember that first day, walking through the halls of Loretto Heights College, my hands full of papers, stopping a woman walking by and asking her where I was supposed to go next.

She smiled, deeply amused, and said, "You must be in the university-without-walls program."

"Yes," I replied, surprised.

She reached out and took the papers out of my hands, sorted through them, and held one of them out to me and said, "Here, this is where you need to go next." She laughed quietly, "None of you university-without-walls students ever read the instructions, every one of you just sort of throws yourself into life and tries to find your way as you go." Then she walked off, chuckling to herself.

"Where I was supposed to go next" was the office of my degree advisor, the person who would oversee the program I was going to create out of nothing more than the urgings of my heart and the curiosity of my mind. We would, together, refine and focus my degree, determine the people I would work/study with, and decide on the subjects I would read, that needed to be part of the degree program. (My degree, since you ask, is in philosophy, in a sub-field called transcultural epistemology, the study of how people in different cultures know and describe reality and the commonalities between those ways of knowing.)

My degree advisor's body filled her chair just as her warmth and intelligence filled the room. She went over the written materials I'd brought with me, listened to me talk for a while, then an odd look came over her face. With an expression full of secret knowledge she said, "Have you ever heard Robert Bly speak?"

"No," I replied.

"Well, you should." She handed me a flyer from the top of her desk, said, "He's speaking in Boulder this week. I think you should go. He fits right in with what you are wanting to do. I think it will help in the formation of your degree program."

I went to hear Robert Bly speak that week and I continued to do so for many years afterward – every time he came through Boulder, in fact. He became, and still is, one of my most important mentors. Importantly, he was the kind of elder for which I had been searching my entire life. As he spoke, I could feel some sort of invisible essence flowing out of him and passing into me. That essence interblended with his words, giving them a particular shape and substance as they emerged from wherever it is that words live when they're resting. In consequence, they became a food that my heart and soul needed in order to truly come alive.

That substance, I now know, has been passed down among us for as long as people have been. It is passed from one generation to the next, from an elder to the young, in a great, never-ending, relay race of soul. I never could have become what I have become without that invisible food filling me, bringing life to the parts of me that had been starving.

I came to understand during those years of hearing him speak that an elder is not simply a person who has grown old. Elders are people who have learned the importance of entering the darkness that surrounds and is inextricably interwoven with the light that we love. Elders have learned to not be undone by their fear of the dark – or the isolation they find there. They have learned, and accepted, that there are truths there that can never be found in the light, that can only be found in the darkness of the world – and in the darker regions of the human heart. And they have decided to enter the darkness, allowing it to become their teacher.

We live, as Robert Bly once said, every day of our lives partly in the land of the dead. Most of the things we take for granted were made by the dead – obvious things like the houses and books and dreams that surround and hold us, but also less obvious things like the soil and the plants and even our bodies (which are given form and life by genes from the dead flowing through time and space into new generations), which we never tend to think of in this way. As he once said, "We receive at birth the residual remains of a billion lives before us." Every morning, though we do not know it, when we wake up, we break bread with the dead.

Elders, because they have traveled more fully in the land of the dead, understand grief in a way that most of us do not when we are young. Over time, what our elders find in that landscape integrates itself into the fabric of their lives. It leavens joy, evens out the turmoil of anger and rage, lessens terror and fear. It brings gravitas and a grounding and solidity that can be found no other way. It is difficult to be whole, I have learned – at great cost – if grief is not deeply integrated into life. Elders, by carrying that integrated knowledge within them, bring to each and every one of us who encounters them (or their words) an experience of something essential to our humanness.

It's called hope – and hope, true hope, is very different than what most people usually mean when the word is used. What is nearly always meant with the word "hope" is, more accurately, optimism – and that is a very different thing indeed. It was optimism and not hope that was the last occupant of Pandora's box. That's why it belonged in the box – and why it was last. It is the same in its nature, but far worse in its effects, than all the other ailments that plague human kind. Optimism is the desire for safety, for things to turn out the way we wish them to so that we continue feel good and safe and valuable and warm and protected. Too often, optimism becomes a way to avoid seeing what is right in front of us. Quite often it is the concealed desire for innocence to continue.

While optimism is one of the essential and (quite often) important attributes of youth, the more one ages, the more important, even crucial, it becomes to grow out of it, to mature into something else. What takes its place is hope. And hope is not and never has been optimism.

Hope is a quiet, enduring, persistent thing. It is not filled with the excited, uplifted, future-oriented energy of optimism. It possesses instead a slow-moving groundedness, an enduringness, a solidity, a nowness. It isn't going anywhere, it just is. It's a form of faith, a faith that comes from life itself. It is faith *in* life and it has nothing to do with personal outcomes or safety . . . and I will talk about it more before the end.

In Bly's presence and words, as I've said, I found a kind of food for my soul for which I'd long hungered. I needed it in order to begin healing some of the terrible wounds inside me. But beyond this I knew that whatever that substance was which he was giving off as he spoke, it was fundamental to my own work and way of life. Somehow, it was core to what I was meant to do. So, I studied his writings (which are extensive) and went to hear him speak every time he came to town, and thought long and deep on everything I saw and heard and read.

By the time I first heard him, he'd been doing this kind of public performance for decades. But, over time, I became curious as to what he was like when he began. So I looked for and found reports of the first, tentative beginnings of his writing and speaking. In so doing I learned a great deal about the journey that all of us take as we struggle to become ourselves. (For when I myself began, I believed that it was only I who had ever been this lost, gawky, uncertain.)

I found early photographs of him, from the time he first started to read his and others poetry aloud. He stands before the room, hair short and carefully controlled, his body wooden, stiff in a dark suit, white shirt, and thin black tie. He holds a book awkwardly in his hands. Those who heard him at the time say his speech was stilted, uneven, the sound patterns of the poems lost in his hesitation and discomfort before the crowd. He rarely smiled or expressed warmth and many of those in the small crowds who came to hear him weren't sure afterward why they'd bothered.

Bly was born in minnesota in 1926, a descendant of lutheran, norwegian immigrants and farmers. They are the kind of people who never use two words when one will do; they prefer to not use words at all. Their particular form of monotheistic protestantism is oppressive, not given to emotions or feelings of any sort; it's especially oppressive in the men. And sex is of course a taboo subject, its energy fiercely suppressed.

Like the men around him, Bly grew up repressed, unexpressive, wooden, relatively emotionless, his sexuality carefully hidden deep inside. Yet, in him burned a fire that would not let him rest. And that fire was a pressure that drove him on, year after year, to face his fear, forcing him to learn how to live, not only in the public eye but in his private body and life.

As time passed, and with more experience, he began to loosen up. He kept the pants and shirt but substituted a mexican poncho for the suit and tie. As it is with all of us when we first try to escape our upbringing, he looked somewhat incongruent, perhaps a bit ridiculous. Nevertheless, he was following the scent of something intangible to the reductive mind and suppressed heart. In later years he found a name for what he was following – he called it a golden thread (which came to him from William Stafford who himself had it from William Blake). As he immersed himself deeper into that thread of meaning, he began to take on more and more of its nature. By the time I first met him it shone out of him, like morning sun after the darkest night any of us can know.

During those years Bly discovered, translated, and brought to the western world some of the greatest poets our species has ever known:

Machado, Lorca, Jimenez, Trakl, Transtromer, Neruda, Vallejo, Mirabai, Kabir, Wright, Duffy, and a great many more whose names escape me now. Because of him words filled with substances american souls needed to find came into our world, words like these by Machado . . .

> *People possess four things*
> *that are no good at sea:*
> *anchor, rudder, oars,*
> *and the fear of going down.*

Or these by Jimenez . . .

> *I have a feeling that my boat*
> *has struck, down there in the depths,*
> *against a great thing.*
> *And nothing*
> *happens! Nothing . . . Silence . . . Waves . . .*
>
> *– Nothing happens? Or has everything happened,*
> *and are we standing now, quietly, in the new life?*

Or these, by Bly himself . . .

> *The dreamer, falling, is about to hit the earth,*
> *and the energy slips him sideways and flows away*
> *with him over the sea, and turns the sword into a*
> *transparent substance that can hurt no one,*
> *and allows a single hair to stir the sea.*

By the time I first saw Bly he'd given up the poncho and was wearing his trademark colorful vests. His body, filled with a childlike joy, exuded life. There was a feeling he gave off as he spoke, almost as if he were dancing someplace deep inside himself. He wore comfortable pants, a billowy white shirt and dark cravat, that flowery and colorful vest – the clothing he would teach in for the rest of his life. Though still filled with the minnesota vowels and consonants of his childhood, his voice was resonant; it filled the room. He would say a line, moving his hands as if he were weaving the meanings in the line into an invisible fabric that only he could sense. When he finished, he'd cast it out over the audience. Then he'd wait, let it settle itself around us, letting the meanings resonate in the air – and inside us – everything immersed in a silence that was itself filled with meaning, as if silence itself was a word filled with import. He stayed with it, never hurrying, letting the meanings reverberate in the room – and in us – as long as they needed, until they reached maximum strength. Then he would go on, maybe saying something like, "That's a nice line, isn't it?"

Despite being so deeply embedded in the woodenness of my life, I sensed that first evening that his words were filled with substances more important than any that had touched me before. They were filled with meanings, and learnings, important to *how* I was then living, the path I was struggling to follow, and the work I was meant to do in this life. It was as if some intelligent force had heard the secret hungers of my heart, the insistent urgings in the soul of me, and somehow, in the inexplicable way that sometimes happens, brought me to this room and those words and that teacher at exactly the moment I most needed to find them.

Bly's memory was phenomenal; without notes he could recite forty or more poems over a two-hour evening, all of them woven into whatever complex, overriding theme he was interested in at the time – which he would tease out in the most delicate and delicious of language. Then he'd illustrate whatever point he was making in the moment with other poems, most by poets I'd never heard of.

As was often true then, that night he was working extensively with the poems of Kabir, the fifteenth-century ecstatic poet. At one point he was playing with one particular poem, part of which goes like this:

> *The truth is you turned away from yourself,*
> *and decided to go into the dark alone.*
> *Now you are entangled up in others, and have forgotten what you once knew,*
> *and that's why everything you do has some weird failure in it.*

I didn't like the poem very much, I found it a bit frightening; I had an uncomfortable feeling that it was talking about me. (It took years before it no longer frightened me, before I could accept that it was, in fact, talking to and about me, to and about all of us who came after as Kabir had intended it to do all those long years ago.)

At the end Bly paused a moment, then said, "You won't be able to understand all this until you're at least forty."

I was offended. What an asshole! (Why had I liked this guy?) I'm pretty smart and really dedicated in my work and my learning; I was entirely confident that he was completely wrong. I could understand this anytime I wanted to. So I worked really, really hard at understanding what he'd been talking about. And then, about fifteen years later, the deeper meanings inside what he'd been speaking of that evening suddenly burst upon me.

How irritating.

CHAPTER FOUR-THE-FIRST-PART

Inevitability and Descent

Complex systems like ecological food webs, the brain, and the climate all give off a characteristic signal when disaster is around the corner.

NATALIE WOLCHOVER

When you haven't been in the world long, it's hard to comprehend what disasters are at the origin of a sense of disaster.

LUCIA BENAVIDES

We live in wooden buildings made of two-by-fours, making the landscape nervous for a hundred miles.

ROBERT BLY

As I sit at my desk, writing these words, my hand, of its own accord, reaches out and runs its fingers over a piece of wood that lies, as it has done for many years, near my computer keyboard. It's a piece of wood from Jacques Cousteau's ship, *Calypso*. As I feel its weathered surface, a scene emerges in my mind from a television interview I saw some forty years ago, in the early 1980s.

Cousteau and a young reporter are aboard a boat on the mediterranean sea; she's asking the usual questions, the ones he'd answered a thousand times before, about how he'd gotten involved in ocean ecology, what drove him to create the Cousteau Society.

The mediterranean sea, in the years before Cousteau fought in World War II, had been a lush habitat, the ecosystem rich and diverse but years later, when he revisited, he found a desert beneath its surface. The shock was so great he'd felt an overwhelming need to respond. The Cousteau Society was part of that response.

Then the woman asks, "Well, do you think you will win?"

There are times which occur for each of us, when something from the heart of the world touches us, and a part of the self, often asleep, awakens. The ancient athenians called this awakening *notitia*, the attentive noticing of the soul. I could tell something important was about to happen, so I began to watch and listen carefully.

For a moment, Cousteau seems unable to make sense of the question, as if the woman is speaking a language he can't quite comprehend. Then, with a very strange look on his face, he says, "One doesn't do it because one thinks one can win. One does it because one must."

He says it quietly, as if it is the simplest thing, as if he is mentioning the weather or what he would like for dinner. He isn't enraged by the environmental devastation of the ocean. He isn't perceiving Earth as victim, himself as some sort of rescuer, fighting against those who are destroying what he loves. He's calm, patient, enduring. He'd obviously felt the impact of all that damage to the life of the sea. What he'd felt was clearly strong enough to motivate him to the work he became known for. And I could tell as I watched that he still felt the pain of the damage done to the oceans of the world – and everything that lives within them. But somehow, he'd integrated it, found balance in the midst of the storm. He could remain himself *without* suppressing the pain. He just *was*. I knew even then, in the midst of my young life, that one day I would need to understand the state of mind and heart he'd found and so I wrapped it up carefully in my heart cloth and have kept it close ever since.

I have not found it an easy thing to achieve the state of mind I saw in Cousteau that day; it's not natural to my character. Yet . . . I sometimes think (though I don't know for sure) that it might be just as hard for every person who struggles to achieve it. For it demands one of the longest and most difficult journeys a human being can take – the one that travels from the optimism of youth (and its desire for safety) to the faith and maturity of hope. And that journey, because of its nature, necessitates fully encountering the pain of the world and the terrible tragedy of the human condition – then integrating it into the fabric of the self as an irremovable aspect of life. It must be done without suppressing the pain while *at the same time* retaining what is most human in us. It means incorporating the inescapable reality of human cruelty, of endings, and of the pain that comes with them into the deepest foundations of the self while still retaining the capacity to love the outward world and an inward awareness of the living reality of the good. No matter what happens as we travel this path we must still be capable of opening the self, and the heart, to the touch of the world upon us.

> *This journey into the pain of the world has taught me a great many things, among them is this truth: the degree to which a person is a knee-jerk skeptic or an emotional cynic is in direct proportion to the amount of pain they have not been able to resolve; it is a measure of the emotional and soul pain they live with every day of their life. Further: each of us who encounters the pain of the world in all its fullness is, at the same time, being asked a crucial question: what kind of person will you become now?*

It has taken years for me to answer that question and I cannot give you a sense of its nature in a single sentence or paragraph or even a book, no matter how elegant they might be. For the answer can only be seen in the fabric of a lived life, in every word that is spoken, every movement of hand and eye, in the ripples that behaviors send into the world, in effects that might not be seen for decades. While it is a question that each of us is asked, the answering of it never ends. It begins anew every day of my life.

For this journey is an experiential one not intellectual; it's neither academic nor rhetorical. The questions the world asks of the deep self – just like the ones asked of us by our souls – are questions that can never be answered easily. For upon those answers the shape of our entire life and self depends. As such they necessitate long contemplation, inner questioning, and years of work. They are the most important questions of our lives.

To my dismay, I discovered as the years progressed that the state of mind I saw in Cousteau that day is not composed of a single insight that once found can be incorporated into the self. Its elements are found in shadows, in things that most of us hide in darkness and refuse to see, in the deepest recesses of the human heart. And it turned out that unbeknownst to me I had been running from them for most of my life.

I remember the first time that was revealed to me. I was in my early forties and, by that time, I'd been teaching for many years. I didn't know why but for some reason my work wasn't going too well. Everything I did had some strange failure in it.

On the day the world changed, I was sitting with one of my students. Oddly enough I don't remember where we were though I can still see the room clearly, as clearly as I see my hands typing these words. We are sitting on comfortable couches and the sun is coming into the room in the way it does sometimes, a honey-golden radiance filled with warmth and the touch of a loving grandmother's hands, the kind of touch that says you're loved and wanted for what you bring to the world.

My student was exasperated with me. She'd been trying to tell me something for a while now and I just wasn't getting it. Then, suddenly, in the middle of something she was saying, she stopped and looked at me with this strange, shock-of-sudden-realization expression on her face and said, "You're afraid. Underneath everything you do is fear. Everything you've been teaching is filled with it." And for some reason, perhaps because I loved who she was or maybe because it was just time for me to hear it, the truth of what she said penetrated through the shell I'd built around me. And my life began to change.

I think maybe it's true that the most important things we hide from ourselves are in a darkness that's as far from the light as we can get them. They are entangled in deep, primal, survival drives that we don't understand . . . or even know we have inside us. There's some kind of strange fear that accompanies them, a belief that if what we are hiding in those dark rooms comes to light something terrible will happen. I have never found the right words for *what* will happen – but I know how it feels.

It's something deeper than words, so deep, it's . . . I don't know . . . something like . . . if I don't keep it hidden, if those things are seen by others, even if the other who sees it is the me who lives up here in daylight, I won't be worthy enough to be allowed to live or to be loved or have enough food or air or . . . something . . . something . . . like that . . . but without words, some sort of existential terror at the core of me.

And it *has* to be kept hidden and there's a part of me that will do anything to make sure it stays hidden and it's been doing that without me knowing it for a hundred thousand years. But . . . there was always that strange sense of failure. And no matter what I did, I couldn't seem to escape it, couldn't figure out where it was coming from or why it was there. Some part of me was trying to tell me that I had a problem I needed to face. But I couldn't even let myself know it.

> *And it's taken years but I've finally come to understand that a third source of the guilt and shame in me has been this cowardice. Some more honorable part of me knew I was running, always running away. And I know this now because when I finally turned and faced the fear and came to terms with it, another chunk of the guilt and shame finally just . . . stopped. And I wonder now, as I write this for you, as I think of all this once more, if all the guilt and shame I've felt, and sometimes still feel, in my life have not always been entangled in some kind of running away from truths I just didn't want to see or hear or know.*

Sometimes, if we are blessed – and it's usually when we least expect it – an event occurs which reveals what we've been hiding from. Maybe a friend or a student suddenly turns on a light and it casts a shadow on the wall and we see a part of our self that we've kept in darkness. The shape that's revealed is bestial, regressed, terrifying . . . and terrified. And for some reason that day, rather than becoming defensive and trying to hide it once more, I turned toward what had been revealed and began the difficult work of facing it and struggling to understand its nature.

It took a long time to understand what my student had been trying to tell me. For as is often true with these kinds of things a large part of me did not want to understand it. There was still, as unfortunately there often is, a great deal of interior work to be done. And interior work, by its nature, especially if there are truths we really don't want to see, is often very difficult. (Or as Ram Dass once put it, "When you first start doing inner work, it's a mob scene in there.")

In-sights are important things but they don't really amount to much unless the work needed to understand and then integrate them is done. Far too often, it's an irritating, painful, goes-on-far-too-long process. Over time, I found (to my dismay) that without some sort of contemplative practice in which the difficult interior work is done insights generally remain only pretty-colored smoke that dissipates on the next windy day. (And, just so you know, a contemplative practice can be anything, from collecting stamps to more a formal sitting meditation to spending long hours letting wood tell you what it wants to be as you shape it into form.)

Eventually, I did come to understand what my student was talking about, did understand the fear underneath and inside everything I did. Oddly enough I found that the fear running as a constant thread through my life and work also tends to be a common problem among Earth activists, among healers of whatever sort. And for me, there were two parts to it and they go like this . . .

The first part is the easiest to see. It's a common attribute in people who are somewhat neurotic, who are introverted, who have thin boundaries between themselves and the world around them – as is true for many of us who are called to Earth work. For lots of complicated reasons that really aren't important to this story I believed I had to earn the right to be alive, to prove that I was worthy of life by doing good works, specifically by teaching about our human interrelationship with Earth, our human responsibilities to Earth, *and* by healing others. And a part of that, though it took much longer to see it, was that I felt a primal need to heal others who were damaged in the same way I was damaged. Like many healers, I was trying to heal the parts of me that had been hurt by healing that hurt, out there, in the world, in other people.

And I think now that maybe this extended to my work healing the damaged ecosystems I so often encountered in the wildness of the world. For we, ourselves, are ecosystems, too, ecosystems that emerged in human form from complex Earth innovation over incredibly long time lines. And maybe some of the wounds inside us have their counterparts in wounds out there in Earth itself. Because . . . what if our interior wounds are merely mirrors for damaged ecosystems out there? What if the wounds inside us don't only come from something our families or we or other people did to us? What if every time an ecosystem or forest is damaged a similar damage immediately appears inside human beings? A reflection of the macro appearing in the mirror of our micro. What if things are not nearly so simple as they've been made out to be? What if there is no real separation between us and the landscapes of which we are a part? And more . . . what if the climate of mind we have been trained to carry inside us damages our interior landscapes in just the same way it damages the ones outside of us? What if, all unknowing, we have, all these years, been clearcutting our internal forests, paving over the wildness inside us? If we could see our interior world in the same way we see the outer world, would we see the same devastated landscapes inside us that we see outside us?

Hidden in those shadows was the belief that I had to actively work to heal others or I wouldn't be worthy of life or love – nor would the part of me that was so terribly damaged ever be healed. And that terrified me. There *was* a fear that lay at the root of my work. And yes, it was in everything I did. And yes, it affected everything I was doing, saying, and teaching. But that wasn't the worst of it. There was another fear, interwoven with the first, but hidden even more deeply in darkness. It was far harder to get to, much harder to see.

It's something that I think lies deep inside every human being who's called to heal. To be clear here . . . not all of those who work as healers are called to the work. Many do it because they need the money or want the social prestige that comes with it or they don't really know what else to do with their life and thank god their parents did and told them what it was even if they were wrong and why do I feel so empty all the time? But those who are *called* to be healers actually *feel* the pain of the world. They feel the pain in damaged people and ecosystems, in Earth itself. And they want to do something about it. That's the healthy part of it. But, as seems to be true with all things, there's a shadow side to this and it entangles itself quite nicely with the drive to be worthy and heal the broken parts of the self.

That shadow side is the strange, nearly wordless drive to heal all the suffering in the world so that it will no longer have to be felt – ever. (It is, I think, the source of all utopian drives; for sure, this was true for mine.) Somewhere inside me, I *knew* how pervasive pain is in the world is (even if I would not let myself consciously know it) . . . and I was terrified of it. I did not want to feel that pain, did not want it to be a truth of this world, of life – of my life. And so, I covered it over, hid it from my conscious self. And throughout the years, I worked endlessly, frantically, doing anything and everything I could to heal *all* the pain of the world so I would not have to feel it. Ever.

Into that frantic drive was interwoven a great many beliefs that are not true – that safety exists, that a life without pain and suffering is possible, that unpredictability can be controlled, that it is possible (*if* they are taught how to do it) for all people to behave honorably and with good intent, that if people truly understood themselves and behaved rationally (or maybe went through their own therapy or extensive schooling – but only the right therapy or schooling and I can tell you what it is) they would no longer do mean or destructive things, like damaging their children or harming Earth or becoming an alcoholic or a drug addict or defrauding people of their life savings or passing laws to disenfranchise others or leaving the bathrooms at gas stations so dirty I feel sick just going into them and why do they always leave the seat down and pee on it and there's so much more and it just goes on and on and on. But underneath all of that, for me and I think for many others, is *fear* of the pain and suffering that is inextricably interwoven with the world. And all those inaccurate, utopianistic beliefs (which are *always* totalitarian if you really think about it), the ones that so many of us erroneously hold – and the insistent, driven work that I continually did – are what I used to keep that fear hidden, to keep it suppressed, to cover it over.

These are the fears that I had all unknowingly interwoven with my work and teaching; they were present in everything I did. To come to terms with them I was going to have to face a great many truths about human beings and myself that I had no wish to see . . . or to feel. I was going to have to learn to accept *as a reality* the way the human world and the people within it *actually* are . . . in all their complexity, their cruelties, their unwillingness to be anything other than what and how they are *right now* in this minute in time. And maybe because Earth loves me more than I will ever know, it was not long after my student confronted me that the world began to show me the reality of what I was hiding from.

Jails are dreary places; I've never much liked being in them. The taupe-colored cement-block walls, the drab floor tile, the smells of lysol and urine and vomit . . . and too, always, the deeper, more insistent smells of fear and testosterone and bravado and ancient, never-ending grief. And those empty, dead cop eyes everywhere you look, the feeling of being treated as a thing with no inherent dignity, the existential loneliness that's seeped into every room and that every one of us in jail feels even if we don't know what it is that we're feeling.

Cops learn to do this thing with their eyes, I suppose they have to but I've always wondered what it does to them after years and years and years of it. Many of the people cops arrest are just, well, average, everyday people. (Though yes, of course others, a smaller but still substantial minority, are not.) They're people who've made a mistake or run afoul of some law they didn't understand or who lost their temper as people tend to do from time to time, or who were broke and desperate and didn't know what else to do. When they're arrested, some lost and fearful child looks outward and into the eyes of another human being, one that just happens to be a cop. What they find, however, is not another human being but "cop" eyes looking back at them.

Cop eyes are created over time; it just takes practice and intent. A barrier is erected right behind the seeing eye. It deflects the gaze of the people who look at them and that is its purpose. It shunts every incoming, emotion-filled gaze off to the side someplace so that it can't reach the heart, can't activate any tendency for compassion or empathy that the cop might still have inside them. It's very difficult to lock up average, everyday people whose only crime is to have made the kind of mistake all of us make in our lives. It's hard to put them inside the kind of inhuman place that jails are if you still feel the reality of what you're doing. To emotionally survive it's necessary to stop all feeling responses to the reality of it.

And so, inside every jail there is an empty, pervasive dimension of being that is ever present. It penetrates the self with the quiet, patient insistence of rust eating away iron. It's the bridgeless space lying between two different kinds of human beings. At its best there is no emotional tone to it – it simply is. At its worst it's filled with the worst of what humans can be. Irrespective of its emotional tone, the arrestees no longer matter as human beings. They are merely an infraction attached to a body which has no interior reality of worth as long as it is contained in that place. . . . And now, as I remember these things another memory, of its own accord, emerges inside me . . .

> *It is a prison in norway but it doesn't look like one. The buildings are aesthetically crafted; there's a homelike feeling to them. They're set in the midst of a healthy, green forest. There's a lake's in the background. There are walls but not like any prison walls I have ever seen. They were envisioned and crafted by a sculptor, an artisan. They are not linear, cement blocks but flow, almost as if they are made of some sort of metallic fabric, in and out of the trees. There is no razor wire on top of them. Nor are there gates or locks or watchtowers. There's openings everywhere. The walls are merely a reminder to the inmates of their separation from the human community, nothing more.*
>
> *The inmates there have committed all manner of crimes, murder and rape among them. Irrespective of this, they work, without supervision, in the kitchen for instance, using, of*

course, very sharp knives to prepare the food. (Many of the inmates are training as elite chefs.) The prison has few if any incidences of inmate-on-inmate violence or of inmate violence against the warders. Recidivism rates are very low.

An american reporter visiting the prison was predictably shocked. She asked the warden how they could even think to do this with people who had committed such crimes. He replied, "We are not in the business of turning men into animals but rather in the business of helping broken and damaged human beings become whole again. But more than that, we are acutely aware of what turning men into animals does to the ones who do it. We do not wish to do that to ourselves or to our nation."

This memory lives as an ever-present question inside me. I have contemplated it – and its meanings – for decades. It causes me to wonder just what other assumptions I have accepted and that live inside me that I have never questioned or thought about deeply. What if I am wrong about more things than I know?

The jail I am in today is different from any I have experienced before. It's new, perhaps a year old. They've already booked and fingerprinted me; I will be released on my own recognizance as soon as the paperwork is finished. After intake, a cop takes me to what I guess is the general holding area; it's where I'm now sitting. Yet it's unlike any I've seen before. It's very large, octagonal in shape, perhaps sixty feet in diameter. There are seating areas, much like those in airports, with long rows of connected seats bolted to the floor. Each section can hold thirty or forty people. There are perhaps four or five of them scattered around the room. The walls are painted in soothing, well-integrated pastels. And except for a strip walkway around the perimeter composed of vinyl tile, it's all carpeted in a very nice, gray industrial carpet.

In the center of the room is a large, octagonal enclosure (the cops' work desk), with a single entryway into it. It has a formica top (light gray), red oak trim running around the edges. Underneath is a quite pretty oak base. There are phones on the counter, computers and screens, in-and-out trays full of papers of various sorts. Inside the desk enclosure stand several cops, among them an extremely thin, almost anorexic woman. All of them are in cop uniforms. The woman is rather short, her face thin and a bit pinched, hair a dirty blond. I can tell she's not been a cop long. She's scared, mostly I think because she doesn't really know how to be or what to do – though she's doing her best to hide it. There isn't a con on the planet who couldn't see through her facade and some part of her knows it, which, of course, increases her fear.

I watch a man, perhaps thirty-five years old, get up from one of the chairs near me and walk up to her and ask a question. She visibly puffs herself up, as if prickles are standing out all over her body, points her finger at him and yells, "Did anyone say you could get up? Sit down. No, shut up, sit down, go back to your seat." The other cops look at her but don't say anything. I can tell from their expressions that they also see what I have noticed in her. Either she'll learn or she won't; time will tell. The man she's yelled at hunches his head lower into his shoulders, turns, and shuffles back to his seat, obviously cowed. He's just a guy, some headed-toward-middle-age suburbanite, his first time in jail, afraid and uncertain.

A guy near me has been watching the whole encounter as well and laughs. I turn to him and we talk a bit. He's in for his second DUI. He's waiting for his pass so he can spend the weekend with his family. He sleeps here most of the time though. He goes to work in the mornings, reports back every night, and if he does well, he gets a weekend at home every now and then. He's got six months to go. He seems comfortable with it all, unashamed. It's just what it is, one of life's problems that, after awhile, will become only memory.

He asks what I am doing here. I tell him felonious mischief. He looks at me oddly, says, "What the hell is that?" And so I tell him. (A few months later the charges against me are dismissed. As the judge in a similar case once said – and as my lawyer told me as well, "While it *is* legal in this state for a citizen, on his own initiative to remediate a public nuisance without assistance from law enforcement, the act is fraught with legal peril. I recommend that you keep this in mind in the future." It's actually pretty good advice.)

There's a phone on the wall, just across the walkway from the chairs in which I am sitting. A woman is talking on it. She's crying. She's latina, obviously dressed for court; she looks quite pretty. She's saying, "I don't know, I just wrote the date down wrong. It was yesterday. When I got here they told me there's a bench warrant and they arrested me. Can you keep my daughter until I get out? Tell her I love her and I'll be home soon. Will you call my boss and tell him what happened? I can't lose that job, I just can't." She cries harder then, sobbing into the phone. She leans against the wall as if it's the only support she has left in her world.

It's then that a door on the other side of the room opens. Two huge cops walk in. Between them are two heavily muscled guys. They're shackled wrists and ankles, chains running from the shackles to a leather belt around their waist. Another chain shackles the two of them together.

They walk with that sort of shuffling motion that ankle chains force upon people. Their faces are slabs of muscle, all sharp planes and angles, empty of expression. Their eyes black, dark, empty holes. Every so often something looks out of their eyes and takes in the room, takes in all of us. Then it slides back into the shadows again. Every time it emerges and runs its dark fingers over me, I shudder and the world turns cold.

I don't know how else to describe it but there's an energy flowing off the men, invisible to the eye yet completely tangible to every one of us in the room. Everyone in the room, and I mean *everyone*, stops breathing. The woman cop's face pales, turns almost paper-white, and she leans, almost falls, against the counter.

We avert our eyes, making sure we make no eye contact with the men. We're frozen in place, all body movements stopped. My breath, like everyone's, is shallow and rapid in my chest, as silent, as nonexistent, as I can make it. Except for the two cops and the men who walk between them there is no movement in the room at all. The cops put the guys in two single, adjacent seats along the walkway and chain them to a bolt in the floor. Then they leave.

Yet we still remain unmoving. For the first time in my life I am in the presence of true human predators. And that energy flowing off the men? It's the feeling of *predator*, filling every corner of the room. Interblended with it is a feeling of evil. It's a terrible coldness that seeps into each one of us. For these men are not merely predators, they are predators on their own species. Predators without moral conscience. And there is not an ounce of compassion in them for anyone.

I don't need anyone to tell me this, no one in the room does. All previous thoughts about my toughness, any sense of myself as strong or as a warrior, dissipate. I know in that moment just how domesticated I am, how domesticated everyone in that room is, including the cops.

These men are what the spartan warriors, the roman gladiators and legionnaires, were like, what human hunter/predators are, why we are one of the few species on the planet that is not endangered. But most striking of all is that feeling flowing off the surface of their bodies. I and everyone in that room knows without doubt that we are being stalked. That if those men were loose, even for a moment, they would use us as they wished to use us, then they'd kill us, and would feel nothing as they did. Some ancient, animal part of me knows this. It is an awareness that's been passed down through a hundred thousand generations of ancestral survivors. To men such as these we are prey, meat, doomed.

I knew in that moment, as clearly as I know anything, that none of the people in this room should ever be in prison with those kinds of men. And I could not help but think of all the fourteen- and fifteen- and sixteen-year-old boys sent to prisons with men like these, of all the soft suburbanites held in prisons with men like these. And of what happened to them when they were.

These men have been born outside their time; there's no place for them in our world any longer – except perhaps as warriors. Yet the genetic ocean still throws them up, as it always has and always will. Some of them learn to blend in of course (just as others do not), to wear expensive suits, to become adept at civilized language, and become lawyers or judges or corporate heads, politicians or bankers or hedge fund operators. But underneath their civilized surface, they remain what they are: predators.

Those men are a reality that most americans never have an experience of. And there are scores of millions of them throughout the world. And I know as well that the capacity for becoming this kind of person is *in* all of us. It is *in* me and it is *in* you. Our genetic code contains it. Inevitably so. My german berserker ancestors were these kinds of men. The viking raiders that raped and pillaged and from whom some of my genes come were these kinds of men. But because of our schooling and television and the domestication of our western world, most of us have forgotten that such men (and women) exist, forgotten that such a capacity is within each of us, no matter how civilized we look to the outside world.

That moment gave me a glimpse of a reality that irremovably lives within the human species. It's something I had long refused to see simply because I did not want such a truth in my world – or inside me. And in the bubble in which I had lived most of my life, that truth had been carefully excluded, redefined, altered. I found myself at that moment, for the first time, face to face with it, knowingly in the presence of evil. And I was undone. I wanted to do nothing more than run back to my safe world and hide. But what is really true, and what I can no longer hide from, is that in our world . . .

> *There are serial killers and murderers and thieves and people who love to hurt others just for the hell of it and people who don't care about Earth and like killing trees simply because they live longer than people do, people who will kill anything they can catch, people who think only people are intelligent and that we can do anything we want to everything else and there are doctors who drain the blood out of living dogs to see what happens when blood pressure lowers and don't feel anything when they do and there are alcoholics who beat their families and they aren't just men and there are drug addicts, too, and people who work for companies that intentionally make drugs that people get addicted to so the companies and people who run them can get rich and they convince the people who take the drugs that if they take them maybe they will live forever or for sure they'll feel safe and not scared or anxious and then there are people who like to control others and people who believe things that I think are really wrong and dangerous and corporations who treat their employees like slaves and police who kill or imprison innocent people by lying about what they*

found or what the person did and are found not guilty when they get caught for doing it and some of them like to torture or beat confessions out of the innocent and there are politicians who are corrupt and always will be but dress in nice suits and there are ideological fanatics and christians who hate everyone who does not believe what they do and will do anything to make their beliefs the law that everyone has to follow and muslims and hindus and scientists and liberals and conservatives and nihilists who do that too and there are people who make weapons that they know will only be used to kill people and they sell them to everyone they can so they can get rich and there are others who leverage poor people into giving them their money to invest and then steal it from them and never go to jail for it and for a long time the united states was the primary maker of thumbscrews in the world and what the hell else are they for except for torturing people and I met a man once whose hands still showed the marks of thumbscrews and I could not look him in the eye and there are tribal people, in their millions, who have cut the hands off hundreds of thousands of people from other tribes and bashed tiny infants' heads against rocks and never felt anything when they did it and still don't and think we are crazy because we do feel something and there are other people who run companies that dump oil and radioactive wastes wherever they want and still others who intend to harvest all the fish in the ocean until everything is dead and they are not going to stop and there are really stupid people that I don't like and mean ones too and all of these people and even more exist and will always exist and are interwoven with the good and the wise and the ones who struggle to grow and know themselves and the ones who just try to be kind and raise their children well and live in harmony with Earth and there are even people who say they are working for the planet but actually are not because they just want the money or the power or the prestige and the people I am thinking about here are mostly just in the united states but there are hundreds of *countries out there each with different languages and cultures and ways of being and thinking that are completely alien to me, and I don't understand why the russians never smile just as they don't understand why americans always show people their teeth right away and is this a predator thing and all those countries are filled with people like the ones I just described and other kinds I haven't even thought of and none of these people are ever going to go away, not ever, and they cause a lot of pain every day of their lives and that pain is **in** the world and it's never going away and I live in the midst of it just like you and everybody does and somehow I have to come to terms with it and I guess that everyone has to one way or another because those kinds of people and the pain they cause are not going away no matter what utopian beliefs or plans anyone has and things have always been this way and that is just the way it is and so how do I remain a good person and still do this work and be in the presence of all this every day of my life, swimming in these kinds of waters and maybe it's easier to just not know these things and to believe even though it's not true that all those kinds of people can and will change one of these days and be just the way I want them to and there won't be any pain anymore and we will all be happy then and the planet will be fine and maybe then I can be happy and not feel this pain anymore, not ever.*

I have spent my life immersed as much as I could be in the good, surrounded by the good, in love with the good. And all that time I was doing my best to avoid the reality of the bad. (I could not yet understand the inescapable reality that, as Bly once said, "Wherever there is water there is someone drowning.") My avoidance of the bad, of the inextricable reality (and banality) of evil, meant that my work was seriously problematical and it was problematical in a *particular* way. At the simplest, my awareness was constrained. There were things I would not, could not, see or understand. As such any suggestions I made about actions to be taken in the world to alleviate our current ecological problems were bound to be problematical as well and Nick Cave was talking about this exact kind of thing when he said . . .

> *Those songs that speak of love without having within their lines an ache or a sigh are not love songs at all but rather Hate Songs disguised as love songs, and are not to be trusted. These songs deny us our humanness and our God-given right to be sad and the air-waves are littered with them. The love song must resonate with the sursurration of sorrow the tintinnabulation of grief. The writer who refuses to explore the darker regions of the human heart will never be able to write convincingly about the wonder, the magic and the joy of love . . . just as goodness cannot be trusted unless it has breathed the same air as evil.*

"Goodness cannot be trusted unless it has breathed the same air as evil." What a magnificent line that is – and what wisdom there is in it. It indicates a necessity, doesn't it? To step into a different frame of reference, to walk away from the good, and breathe the same air as evil and find out what happens when you do.

And so, that is what I did.

Each and everyone of us sees what we focus on, often to the exclusion of everything else. There is a set-switch inside us that adjusts our sensory gating channels to allow some things in and tune other things out. (We are thinking about buying a toyota 4x4 and for a while we see them everywhere.) At the moment I opened myself to the bad, to the evil that is in our species and which we so often remain ignorant of, for a time that was all I saw and felt. There was pain with it, terrible pain. It's a pain, that, in aggregate, exists, irremovably, *in* the human world. And it has existed ever since human beings have been.

I entered that reality, opened myself emotionally to it, internalized it. I breathed its air, let it dominate my awareness. It became, for a time, the only thing I knew.

I found it terribly difficult to adjust to. For seeing/feeling the world this way is a disheartening thing. Almost of its own accord it stimulates a cynicism, a bitter denunciation of the good as something unsophisticated and naive. And underneath that bitterness and cynicism a terrible pain that is very difficult to endure hides itself.

Nevertheless, I carried on, immersing myself in its textures, learning to see through its eyes. And, once the shock of it wore off enough, I began to find my balance in the midst of it. I began to look around, to make distinctions, to examine the territory, to find my way inside it.

I began to study it, to read about it, to listen to the words of people who had entered its world more fully than I would ever do. People like Viktor Frankl and Vaclav Havel and Black Elk and Hannah Arendt and James Baldwin. They were honorable guides, for they had breathed that world and in its midst found a way to remain themselves, to uphold the best in them.

> *I don't think that people really understand what Frankl means when he says, "The good ones [in the concentration camps] all died," because you know, he survived. In his face, too, I have seen the same look as I did in Cousteau's.*

I struggled to understand that world, with my mind, with my heart, with my feeling self, and with the child who still lives within me, the part of myself that has always approached the world with wonder and curiosity and love. And maybe that was the hardest thing of all to do for the child within us can't easily understand evil.

While all this thinking and feeling and struggling was going on (though I did not know it for the longest time) a unique process was taking place within me. And it turned out to be crucial to the state of mind I was seeking.

During those early days, immersed in those dark waters, while I was solely focused on experiencing the bad, feeling the impact of the evil I had so aggressively ignored, I wasn't thinking any further than that. That dark world became a lens through which I saw, and felt everything. It was the dominant reality of my life. My heart compressed under the weight of it, its inherent cruelty was seemingly everywhere; life itself began to lose meaning. I walked each day as if a shroud of darkness lay over me and nowhere could I find goodness or kindness or joy. I caught myself thinking, "If this is the truth of human life, I don't see how I can go on. Nothing will ever get better, only worse." I knew that I was experiencing an inescapable reality, that it *is* **a** truth of human life, and that it is not going to go away. Not ever.

In the midst of that darkness, my only guides were the words left by those who had themselves traveled here, my sense of the right, and the feeling that, even if I could not feel it now, there was another shore, another way of being somewhere out there in the wildness of the world that maybe I would find if I kept on. There was, too, a sense that this was something I must do.

Eventually, the darkness began to break. I started to remember, and to feel, the good again. For while it is true that the bad is an irremovable reality, so, too, is the good – whether or not I can feel it in any particular moment, on any particular day or week or year. And from time to time I would remember its touch, the feel of it, the feel of love and caring that came to me from those whom I love and who love me. And I'd remember as well that there are people who strive to become more than the bad or the evil that is within them. Somehow they had found their way through this, and if they could do it, then so could I.

Eventually, I began to oscillate, back and forth, from one world to another. The bad would become dominant and I would see the world through that lens alone, feel only that, know only that world as real. Then the good would become dominant and I would see through *that* lens, feel only that, know only that world as real. Two irreconcilable opposites coming and going within me. Over and over and over again.

But one day, of its own accord, the two suddenly and inexplicably joined together and became a unity. I found myself holding *both*, simultaneously. The full reality of *good* and the full reality of *evil* were inside my mind and heart *at the same time*. They took on a unified form then which contained the reality and identity of both yet possessed something more than the sum of their parts. For the first time they became *good/*

bad, a combined *yes/no*. A state beyond either/or came into being inside me. It was then that I first found the state of being and mind that I had seen in Cousteau on that day so long ago. And I will have more to say about it before the end – because this is just the beginning of what can be found in the unification of paradox.

The journey through ecological loss and its grief necessitates an engagement with evil, with the bad that is *in* the human world, which has always been in that world, and always will be in that world. Many of the troubles we face come from men and women smelling of evil, inquiring into the purchase of our homes. They operate from a very different orientation than those of us who love this planet – and what is true and will always be true is that creating a utopian world in which these people do not exist is not possible. It will destroy us all if we try, if we take that road as the solution to our problems. Something else, a different approach, is needed.

To do the work we are here to do it is incumbent upon us to have the courage to breathe the same air as them, to walk in their world, to understand it fully and *at the same time* not lose ourselves in its embrace. We have to learn to walk in the midst of evil and remain ourselves. We have to be brave enough to be more than we have believed we could. And as I write these words I remember something that Aleksander Solzhenitsyn said in *The Gulag Archipelago* long ago . . .

> *Gradually it was disclosed to me that the line separating good and evil passes not through states, nor between classes, nor between political parties either – but right through every human heart – and through all human hearts. This line shifts. Inside us it oscillates with the years. And even within hearts overwhelmed by evil, one small bridgehead of good is retained. And even in the best of hearts, there remains . . . an uprooted small corner of evil. . . . [But] in keeping silent about evil, in burying it so deep within us that no sign of it appears on the surface, we are implanting it and it will rise up a thousand fold in the future.*

In the end I think perhaps my student might have taught me far more than I ever taught her. For not only did I learn about these hidden fears and a way of life that comes after they are faced, but I also learned a truth that all teachers must learn sooner or later: *teaching is not and never has been like a river, it doesn't flow in only one direction.*

Grief will be our companion on this journey, but also pain and loss, for they are the handmaidens of grief. And of the many pains in the world, there are three that have made their home within me. Each has to be faced sooner or later, for they will be encountered over and over again on the journey into and through ecological loss. The first I have already spoken of. It's the pain that comes from our living, from the wounds we receive as we rub up against other people and as they rub up against us. It's the pain of the human tragedy and its never-to-be-given-up cruelties and limitations and blindnesses. It has accompanied our species since our species has been. It's the pain that comes from the cruelties, intentional or otherwise, that our species does to itself.

The second is the pain that comes from our biodegrading, from the inevitable dissolution of all the complex systems that make up our selves and our life: the physical, the emotional, the psychological, and the spiritual falling apart of us as death comes closer, as death arrives. The pain that comes as the "I" that we think we are is slowly subsumed back into the ecological matrix of Earth, as the "I" that every form of life has, and is, is slowly absorbed back into Earth. And that pain is not personal to the human species nor our individual selves. It just is. And it, too, is something that must be faced, and embraced, in just the same way as evil must. For dying and death are going on all around us every minute of every day of our lives (even though we pretend they are not).

We are surrounded, not only by the dying and the dead, but by the pain that every life form on Earth has felt and is feeling right now in this exact moment in time. Those of us who are sensitive to the touch of the world upon us, whose boundaries are somewhat porous, know this, no matter how deeply we have buried that knowing. Many of us have learned, unconsciously I am pretty sure, to modulate how much of that pain we are willing to let in. We feel the pain of other life (well, *some* other life), perhaps that of other people (well, *some* other people), or animals (*some* animals, certainly not cockroaches), or landscapes (or maybe just *some* landscapes).

This, too, is a pain that must be integrated into the self. The process is identical to that of integrating *good/evil* but it deals with a different set of "opposites." It is concerned with the unification of *death/life* and that is an entirely different thing. And it, too, is something I will talk about more before the end. For it is from the unity of life/death that hope comes.

Then there is the third pain. In a sense it can be thought of as an extension of the second pain. But it is unique in our time because it's occurring at a level that has not happened in a very long time, in fact, it has not happened since our species emerged on this planet. Earth is once again entering a geologic age where the ecological damage is so vast that planetary function is destabilizing. In consequence, Earth's ecosystems and life forms are destabilizing, too – falling apart, collapsing, dying. They are dying in their millions, in their billions, in their trillions – and there is not a person (or a life form) on Earth that does not feel it. Dying and death and the pain that attends them is happening at levels not seen in sixty-five million years.

With increasing urgency, over the past half century, our species has been being given a very specific diagnosis. The people who have been giving us the news, so to speak, irrespective of their titles or schooling (or lack of them), are, in an important sense, the ecological physicians of our time. They're the doctor telling the patient that there is a small tumor in the lung and that surgery is needed – and most especially, since the lung has been damaged and cancer is developing they should really stop smoking. (Analogies like these always break down and they're never completely accurate and I don't particularly like this one but it will work well enough for a little while, maybe . . .)

As is always true with a terminal diagnosis, there are a great many people who do not want to believe it, who are sure that it's bullshit no matter what anyone says. There are others who do believe it but refuse to change, because screw it or I don't want to or why bother, it's all hopeless anyway. Others have their own islands and their billions and they don't really care what happens to the rest of us. There are still others who are demanding second and third and fourth opinions. And there are others yet who are sure there's a technological fix out there someplace that will save them so they don't really take the diagnosis all that seriously because *science*!

The diagnosis, however, is a lot worse than it was a half century ago, for the problems before us now are far worse than they were then, when we, as a species, were first told about what we were facing. And I think that it's about time for all of us to look with unafraid eyes at what is right in front of us, begin to seriously grapple with the truth that's been set before us, to actually look with clarity at the diagnosis we've been given. For one of the maxims of life is that if we understand the situation in which we find ourselves it is far more likely that we can take the actions needed to deal with it. And the actions needed are far different than those that most activists are talking about. (Though, yes, of course, *some* of what they are suggesting – imploring, insisting, advocating, demanding, pleading for, telling us we must do or they will shame us endlessly if we do not – can help . . . a bit . . . probably, maybe, I guess.)

These days, the diagnosis is most often oriented around global warming. Occasionally other aspects of our troubles are presented, water depletion in the west perhaps, or the worldwide loss of insect species, or maybe the ecological impacts of industrial food production. But, in actuality, what we hear, what is being printed in the media, is only a tiny part of the reality we are facing. The world's people are aware of less than one percent of what is actually happening to the ecological fabric of this planet (and no, Kim Stanley Robinson's science/fantasy books don't really get it either so don't bother going there). The situation is far worse than most people realize. Global warming is in actuality the least of our problems (though, yes, it is indeed a problem). In fact, all that focus on global warming obscures just how bad the situation really is. (And it covers over the complexity of what needs to be addressed by making the solution appear to be an essentially simple one – the use of a different kind of fuel to power technological civilization.)

Now all this sounds like I am about to get into the usual doom and gloom stuff that is surrounding us every day of our lives now, doesn't it? And yes, I am, kind of, I guess. But the purpose of my talking about the way I am about to is very different than most of the articles you might have read or news reports you might have seen.

It's not to frighten you (though it probably will) or to stimulate you to "do something" (a motivation I particularly loathe). The purpose is something else again. It's to understand the actual diagnosis and what in fact it really means because . . .

NARRATIO INTERRUPTUS

The Diagnosis

Growth for the sake of growth is the ideology of the cancer cell.
                                                    EDWARD ABBEY

We are less than a decade away from carbon dioxide levels reaching 450 parts per million, the equivalent to a 2-degree Celsius average temperature rise, a global catastrophe that will make parts of the earth uninhabitable, flood coastal cities, dramatically reduce crop yields and result in suffering and death for billions of people. This is what is coming and we can't wish it away.
                                                    CHRIS HEDGES

Our epitaph [as a species] may well read: they died of a peculiar strain of reductionism, complicated by a sudden attack of elitism, even though there were ready natural cures close at hand.
                                                    GARY PAUL NABHAN

Great civilizations are not murdered. Instead, they take their own lives.
                                                    ARNOLD TOYNBEE

Although reports of our dire situation make the news every day of our lives few people understand just how pervasive and extensive the impacts of unrestrained corporate industry and technology have been and still are in destabilizing planetary function – nor do they have a sense of how much worse it's going to get. There are hundreds of ecological systems that are failing that I know of, there are thousands more that I don't know anything about. It runs from soil death and depletion . . .

> *Some one hundred million bacteria, yeasts, molds, diatoms, and other microbes live in just one gram of ordinary topsoil [as Masanobu Fukuoka explains]. Far from being dead and inanimate, the soil is teeming with life. These microorganisms do not exist without reason. Each lives for a purpose, struggling, cooperating, and carrying on the cycles of nature.*

. . . which turns soil into unliving dirt, to aberrations in earthworm life and behavior to underwater aquifer depletion and groundwater contamination to the massive worldwide deaths of insects to coral bleaching to the collapse of the ocean's fish populations and ecology to the deep mining of the planet's ocean floors to the worldwide melting of ice to increased hurricane incidence to alterations of global atmosphere and its dynamics to destabilization of global plant populations and behavior to massive die-offs of bird and large mammal populations and the subsequent impacts on soil fertilization and herbivory to disturbances of the planet's microbiome and function. And this is only a very tiny part of the list. *Every* planetary subsystem is destablizing, every part of it is in peril from corporate industry and technology.

While there are many sources of destabilization I am only going to talk about two: plastics (briefly) and the medical industry and its pharmaceuticals (in some depth). I am doing so for two reasons: 1) so you will have a sense of the pervasive ecological damage that only two of the many corporate industries that exist are responsible for (one of which is almost never discussed in the media while the other merits only limited, relatively superficial discussions of its dangers), and 2) just why it is impossible to stop what is happening to the planet's ecological systems.

One thing you need to know up front: both the physician/hospital/pharmaceutical/medical industry complex and plastics are totally and completely dependent on petroleum for their existence. Plastics and the majority of pharmaceuticals are *made* from oil. The oil companies are not all that worried about losing gasoline as revenue; they have long intended to shift their primary focus to plastics and the ever-increasing market for pharmaceuticals. Neither plastics nor pharmaceuticals are going away. And neither is the world's need for oil.

The Ecological Impact of Plastics

Polymers are forever!
Alan Weisman

Ants make plastics, too.
They just don't cover the world with it.
Alice Walker

I recently stumbled across a YouTube video of the interior of a store in rural Canada. It had been in business since the late 1880s, closing in the early 1960s. The elderly owner, a descendant of the founders, had died (while working at the counter as it happens) and her family did not want to take over the store. So, it closed and the contents were left in place where they have remained now for some sixty years.

As the camera panned around the store and its shelves it was soon obvious that there was not one thing made of plastic in the entire store. *Everything* was either glass or metal or wood or paper or wool or made from plants and plant fibers of one sort or another. *Everything* was biodegradable or reusable, *everything* was made from what is now considered to be simple technology. There was very little chemical modification of natural compounds involved.

To get an idea of what it is like now, just go into any store and look around. Nearly everything in the place will have plastic as part of it in one way or another. Plastics are everywhere and in or on everything that people buy, even our food.

> *I remember when the contents of every store in the United States were like the one that I saw on YouTube. It wasn't so very long ago in people years; it's the blink of an eye in Earth years.*

Now, consider America's national obsession with single-use plastic bags. How, because of environmental activists' lobbying they are being "phased out." And all the popular uproar about what they are doing to the environment . . .

> *I remember when environmental activists and tremendous public pressure convinced legislatures to outlaw paper bags, forcing businesses to switch to plastic in order to save the forests from paper bag manufacturing ("An entire forest is cut down every day to make your grocery bags!" the headlines at the time screamed). As usual, the cure people created to "save the environment" was far worse than the disease.*

Nevertheless . . . take a minute, go into your kitchen and look around. You will find that plastic, in whole or part, is included in the manufacture of everything you see . . . okay, not the glassware . . . yeah, yeah, okay, and not the silverware either . . . now stop it.

The hysteria over single-use plastic bags is just a form of ecological theater (just as plastic recycling is). It won't do much about plastic pollution in the long run – Americans throw away 75 *billion* plastic water bottles each and every year, for instance – it just diverts attention from where it should really be going: to the incredibly dangerous ecological impacts that come from the manufacture and use of plastics in any form at all. The thing is, there are millions of plastic products now. They are so ubiquitous that people no longer notice how pervasive they are. They are just a part of life, like coastlines and the sea.

> *And . . . what about all those millions of single-use plastic gloves that are used in the world's hospitals **every day** of the week and are just thrown in the trash? (And the millions more*

> *used during the Covid-19 crisis by hospitals and the general public.) What happens to them? No one talks about that very much, do they?*

What is true is that all of the plastic that is incorporated into your kitchen – *every single bit of it* – when you really think about it, is single-use plastic – despite the fact that some of the products that contain it might be used for a decade or more. (Even the fruit we buy has little tiny plastic stickers on it now.) *None* of that plastic is truly biodegradable. Once those products reach their end use, all of it will end up in landfills. And it will last for a very long time, centuries in fact. (And no, bioplastics are not the droids you are looking for.) As Alan Weisman has so succinctly put it: *Polymers are forever.*

If you then extend your plastic perception outward (say to automobiles, for instance), you will find that plastic is in pretty much everything that is manufactured. It has penetrated every aspect of our lives. The electrical wires in our homes are plastic coated (they used to be wrapped in fabric), the pipes are plastic (they used to be metal), the lamps have plastic turn-on knobs (used to be metal pull chains), computers and printers and televisions are all manufactured around plastic (and shitty plastic at that). Shopping carts are coated in plastic, rugs are plastic, flooring is plastic, windows in houses have plastic frames, doors are often plastic or plastic coated, paint is plastic, clothes are plastic, shoes are plastic, paper clips are plastic coated, pens are plastic, cups are plastic or have plastic tops, eyeglasses are plastic, and all those billions of automobiles that are on the roads of this planet contain an average of 400 pounds of plastic per car (and all of that goes into the trash when those billions of cars no longer work). Everything we buy is wrapped in plastic or comes in a plastic container, pill bottles are plastic, tubing is plastic, gallon jugs are plastic, 5 gallon buckets are plastic, "rubber" tubing and tires are plastic, astroturf is plastic, children's toys are plastic, cell phones are plastic, and on and on and on.

All these things, and a great many more, are being made in their millions, their billions, their trillions for all the billions of people on the planet. And most of the products that contain plastic are intentionally manufactured to be of limited duration. (It's called planned obsolescence.) Corporations long ago gave up making products that would last and work well for decades or even generations. They make far more money if people have to replace everything on a set schedule. Thus they intentionally design a limited life span to what they make and what we buy. In contrast, my grandmother's blender, refrigerator, toaster, and stove, all bought when she married as a young woman, were made of metal and glass. They worked well for a half-century, *fifty* years, and there was very little plastic, if any, on or in them. They were still working when she died. America used to be known for the reliability of its industrial products (Sears, for example, had a lifetime warranty on its tools). The most recent toaster I bought lasted three years. The computer six, the printer five.

Again: none of this plastic I am talking about is biodegradable. And that is a very serious problem. All plastic, one way or another, goes into the environment. Over time it breaks down into what are called microplastics (which you may have heard of) and over time those microplastics break down even further into nanoplastics. Both of them are very, very dangerous.

"Every human on Earth," as Tim Dickinson says in his *Rolling Stone* article "Planet Plastic, "is ingesting nearly 2,000 particles of plastic a week. These tiny pieces enter our unwitting bodies from tap water, food, and even the air, according to an alarming academic study sponsored by the World Wildlife Fund for Nature, dosing us with five grams of plastics, many cut with chemicals linked to cancers, hormone disruption, and developmental delays."

This is not new information; it's been known for a long time – at least to the companies making the plastic. So has the fact that plastics are not easily biodegradable, if at all. (Plastic manufacturers and the oil companies are insisting that while it is true that micro- and nanoplastics affect other living organisms, there is no proof they are harmful to people. This is a specific strategy based on what is called the Kehoe Paradigm, which states that all manufactured chemicals are considered to be benign unless definite harm can be proved. In consequence, the companies continually create scientific uncertainty in order to keep their products from being regulated – as they have done with tobacco and so many other things.)

Human-created plastics are synthetic; they do not exist in nature. (There are a very few naturally generated plastics in the ecosystem. There isn't much of it and it's far different than laboratory-created plastics.) Human-made plastics are unique in Earth history and there are no existent natural systems (from the micro to the macro) that can biodegrade these unusual chemical compounds as they do everything else. While Earth systems will eventually figure out how to do so, it won't be anytime soon.

> *And no, all those techno-utopianistic proclamations (that you see from time to time in the news) that this or that process or microbe can be created or tweaked or modified to biodegrade the world's plastic wastes ("Scientists create bacteria that can turn plastic waste into vanilla! Thus solving the world's plastic problems!") and so make them a non-problem is **never** going to happen. And even if such a thing were created, at most it will only create other, more terrible, and unforseen side effects. Like . . . what happens if they do actually genetically modify a bacteria that can live by eating plastics? What happens if it gets loose in the human world? Or, say, in a plastics factory? Or in the world's computers? Cars? Medical equipment? Clothing? (And no, they won't be able to contain it when it does, haven't you seen the movie?)*

It's going to take centuries for the planet to recycle all the plastic waste which now exists and that will continue to be manufactured for the foreseeable future. Further, it's not just the plastics that the system must deal with. A 10-pound bag of plastic waste generally contains more than a thousand discrete chemical additives. Those have environmental impacts, too, little of it good, few of them studied.

And all that plastic? It begins with what are called nurdles.

Nurdles are tiny plastic beads about the size of a lentil. They are what all plastic products are made from. Nurdles are made by petrochemical plastic manufacturers and then sold to every plastic product maker on the planet. Some of the most powerful corporations on Earth are involved: Exxon, Dow, Shell, Coca-Cola, Nestle, Unilever. (This is business as big as it gets and trillions of dollars are involved – they are never going to give up the money.) These monster corporations make single-use water and soda bottles, single-use plastic packaging, single-use plastic bags, single-use hypodermic syringes, pill bottles, medical tubing, and so on and on and on. As Dickinson puts it, "Big plastic isn't a single entity. It's more like a corporate supergroup: Big Oil meets Big Soda – with a puff of Big Tobacco, responsible for trillions of plastic cigarette butts in the environment every year." (He left out Big Medicine, as is often the case.)

Most of the plastic polluting the world has been manufactured over the past twenty years. Another way to look at this is that in two decades this one chemical product went from virtually no ecological impact to one of the most damaging on the planet. In sort-of-graspable terms, a dump truck load of plastic waste goes into the oceans of our world *every minute of every day of every week, month, and year.* (By 2050 plastic waste in the world's oceans will weigh more than all the fish in the sea.) But as bad as this is, as one researcher observed, "This is a much bigger problem than 'just' an ocean issue, or even a pollution issue. We've found plastic everywhere we've ever looked. It's in the Arctic and the Antarctic and in the middle of the Pacific. It's in the Pyrenees and in the Rockies. It's settling out of the air. It's raining down on us." It has been found on the highest mountains, in the remotest locations, and at the bottom of the Mariana Trench, the deepest region of the sea.

Again, nearly all, and I mean *all*, of the plastics that are made are single use plastics, they are neither biodegradable (at least in any of the next eight generations' lifetimes) nor recyclable. As environmentalist Jim Puckett comments, "They really sold people on the idea that plastics can be recycled because there is a fraction of them that are. It's fraudulent. When you drill down into plastics recycling, you realize it's a myth."

> *I define multiple-use plastics as plastics that are truly recyclable and that can be made into other forms for repeated use after the original product reaches end-of- life – this is a very rare*

occurrence. Most "recyclable" plastics can only be reused once. (And all the plastic waste being used to make asphalt roads? As tires travel over them tiny pieces of both tire plastic and road plastic are released into the environment.) To define a printer as multiple-use ignores the fact that they all end up as waste in the environment where the plastics they are made of will continue to wreak havoc for a very long time to come. Thousands of them are thrown away every day. Just because we use something for a few years rather than one day doesn't make its manufacture sensible or ecologically benign.

In the last twenty years, corporations have created *14 trillion pounds* of plastic waste, 91 percent of which has never been recycled (it can't be). And the small percentage that has been recycled? It can only be recycled once. (As an aside, plastics are so ubiquitous now that they are becoming part of the geological record, being compressed into a distinct stratum in the geologic layers of the planet. They have formed a new kind of rock called plastiglomerates.)

Most of this plastic waste is simply deposited in the world's waters, much of it ending up in the ocean – about 17 billion pounds of it each and every year. (Five hundred thousand to one million tons of it every year is commercial fishing gear lost at sea.)And those numbers will only escalate as gasoline becomes less important to transportation (because plastics and not gasoline is where the companies know their future growth and profits lie).

Only about one percent of the plastic deposited in the ocean ever reaches the shore or surface of the sea. This one percent is what you are seeing in those photographs of ocean plastic waste. *One percent.* The rest remains under the surface, unseen, out of sight, out of mind. In your imagination, increase *by 100 times* the plastic beach and ocean garbage patches that you've seen in photographs – if you can. (I find it difficult to do.) That will give you a sense of just how much plastic is in the ocean. (And this doesn't include all the micro- and nanoplastics in the world's seas – they are not visible to the eye. And there are a great deal more of those than the ones visible to the eye.)

Plastics are not only in the ocean, of course. They are in the world's soils as well because they are extensively used in agriculture: "They are used to wrap silage, to cover crops, in tubing for irrigation, to transport feed and fertilizer." And as Liang Lu et al comment further, "In fact, MPs [i.e., microplastics] contamination in the land may be more serious than in the aquatic environment due to the directly extensive use of agricultural plastic film and the plastic particles in the industrial production." Machado et al note that "microplastic contamination on land might be 4-23-fold larger than in the ocean. Indeed, agricultural soils alone might store more microplastics than oceanic basins."

They are in the soil, in the air, in freshwater lakes, in streams, in arctic ice, in our drinking water. Everywhere. (The filters in water treatment plants, by the way, are also plastic; they degrade from friction during use, adding more micro- and nano-plastics even as they purportedly clean our drinking water.) As Dickinson says, "The pollution is planet wide, impossible to fully remediate, and threatens to disrupt [actually, already is disrupting] natural systems." Or as Ramsperger et al comment in *Science Advances*, corporations "are conducting a singular uncontrolled experiment on a global scale in which billions of metric tons of material will accumulate across all major terrestrial and aquatic ecosystems on the planet."

Despite increasing scientific and public concern, big plastic has no intention of stopping production. As Beth Gardiner comments at *Yale Environment 360*, "While individuals fret over images of oceanic garbage gyres, the fossil fuel and petrochemical industries are pouring billions of dollars into new plants intended to make millions more tons of plastic than they now pump out." As she continues, "Companies like Exxon-Mobil, Shell, Saudi Aramaco are ramping up output of plastic – which is made from oil and gas, and their byproducts – to hedge against the possibility that a serious global response to climate change might reduce demand for their fuels."

Plastic production is expected to make up half of oil demand in the near future and this is only going to increase. To truly correct the problem, we need to, as a species, return to glass and wood and stone and metal and natural fibers – something that is obviously not going to happen. There is simply too much money and power involved to reverse course now. And at this point, too much of the world industry and infrastructure depends on plastic to shift back. Instead we are going to have to live in the midst of more and more plastic waste . . . or perhaps I should more accurately say: plastic waste is going to live in us.

Plastics, when they enter the environment, don't break down, they just become ever smaller as exterior forces have their way with them. (Friction breaks them down, just as it does rocks and shells while creating sand.) The smaller the plastic particles get, the more easily they enter the bodies of living organisms, including us.

Plastic has been found in the stomachs of at least 220 marine species, from the smallest to the largest (over 200 pounds was once found in the stomach of a dead whale). The smaller the waste is, the smaller the organisms who ingest it. Nurdles, for instance, are considered a microplastic due to their small size. They are both clear and colorless (so they can become any color a manufacturer desires). They happen to look very much like fish eggs. And so, when nurdle manufacturers suffer their inevitable spillage events, billions of nurdles go into the world's waterways. They are then eaten by a large, diverse grouping of fish and birds.

Even tinier microplastics are eaten by even smaller organisms: plankton, nematodes, roundworms, springtails, and mites, for instance. Plastic waste is incorporated, one way or another, into the bodies of everything that consumes them and there they remain. Living organisms' bodies have no way to get rid of the plastics they ingest. So they are incorporated into fat and the body's cellular structures. From there they move up the food chain as larger organisms eat smaller ones.

Microplastics are very tiny (less than 5 mm in size) and they have been found in every living organism in which researchers have looked. This includes all plants (which absorb microplastics through their roots which then become part of the plant itself, you know like carrots, apples, and lettuce); algae and phytoplankton (where they interfere with those plants' absorption of sunlight and hence reduce their health and usefulness as food for other animals); yeasts, fungi, and bacteria; mussels, clams – all shellfish; all fish (North America's salmon are dying from it); animal muscle tissue (such as turtles and panthers); most if not all vertebrates; and of course, us.

On average, human beings either eat or breathe in around 50,000 microplastic particles every year. These tiny particles then move rather easily through the lung and GI tract membranes into the body's circulatory system (blood and lymph) and from there deeper into the body, where they become part of our physical selves (Plastics R Us). They've been found in most human organs, the lungs, liver, spleen, testes, ovaries, heart, and kidneys for example; in our muscles; in every part of our bodies that's been examined. These tinier microparticles are easily incorporated into our cells (and, yes, they have been found there), where they exert a wide range of effects.

The intake of both micro- and nanoparticles causes a generalized inflammation throughout the body of every organism that has been studied. These tiny plastic particles interfere with the intestinal barrier in the gut and, as well, alter the microbiome of the entire GI tract. Both micro- and nanoplastics are taken inside cellular tissues, including the bacterial where they alter behavior and microbiome community makeup. (In other words, the healthy ecological function of the human microbiome is disrupted.) In the intestinal tract the plastics cause inflammation and oxidative damage, destruction of the gut epithelium, reduction of the protective mucus layer that lines the GI tract, and immune cell toxicity (in other words, irritable bowel syndrome, Crohn's, and so on).

The plastic causes immune cells throughout the body to disregulate. This begins with the immune responses and cells of the GI tract, moving outward from there. Immune cell counts are reduced, activity decreases, and function is impaired. There is an increase in the production and activity of neutrophil extracellular traps (NETs, which in and of itself causes systemic problems), myeloperoxidase activity, and leukocytes. The complement system is impaired; inflammatory cytokines increase and their activity heightened.

These same impacts are seen in the lungs (the GI tract/lung microbiome is in fact a single system; what happens in Vegas does not stay in Vegas). There is disruption of the lung microbiome, its immune function and responses, its protective "mucus" layer, cilia activity, alveolar and alveolar macrophage activity and function. The microplastics are easily transported from the alveolar space into the blood, where they, again, reach every organ in the body.

The alteration of gut and lung microbiota directly leads to system-wide effects on health, including in the brain and central nervous system. The inflammation that microplastics cause is not limited to the intestinal tract or lungs, it is system wide. In fact, ingestion of microplastics causes a continual low-level inflammation throughout the body, including the brain.

Because microplastics usually contain other toxic chemicals (coloring agents and so on), these chemicals also enter living cells. In essence the plastics act as a carrier for the movement of highly toxic chemicals into all living systems. Further, microplastics in the sea often float on the surface, where they are colonized by a large range of bacteria, including cholera organisms. In other words, plastics act as carriers for microbial pathogens – and they enhance the toxicity of those pathogens. As Liang Lu et al comment, "MPs accumulated by viruses and bacteria are more biotoxic than ordinary MPs. After entering the organism, it is easy to cause organism infection." In other words, the microplastics are altering both microorganism infectivity and behavior.

Nanoparticles are even tinier, less than 0.001 mm in size. And they are pervasive in the human body as well. Nanoparticles have been found in pregnant mothers and their fetuses, in the human brain and central nervous system, inside our cells and inside the microbial members of our microbiome. They have been found in every organism that has been studied, including the microorganisms that are foundational to the functioning of this planet. As Tim Smedley, writing for BBC Future, comments, "The biggest [pollutant] killer of all never makes the headlines, isn't regulated, and is barely talked about beyond niche scientific circles (despite their best efforts to change that narrative); it's nanoparticles."

The impacts on the central nervous system are severe. Nanoparticles easily cross the blood/brain barrier – they also find their way into the brain via olfactory nerve endings, much the way Covid-19 does. Once in the brain they induce oxidative stress and damage its neuronal structures. They affect the mitochondria, thus disrupting the body's energy metabolism, alter acetylcholinesterase activity, and cause a wide range of neurobehavioral impacts. Astrocytes become reactive (astrocytosis). Once reactive, they generate increased levels of lipocalin-2 and proinflammatory cytokines. The nanoparticles also enter the brain's neuronal cells, where they cause shorter neuronal life spans and poorer function. Cleaved caspase-3 is significantly elevated within the brain. Nanoplastics stimulate cellular death, alter neurotransmitter levels, locomotor behavior, spatial recognition memory, cause impairment of learning and memory, and dysregulation of glutamatergic signaling. Neuroinflammation is common.

Similarly to microplastics, nanoplastics cause a system-wide inflammation in every organism that ingests them. They easily enter a wide variety of cells, where they act as genotoxins, that is, they cause genetic damage. They easily move across the placental barrier and into the fetus (often after being breathed in by the mother). As the fetus develops, the nanoparticles affect the development of every organ due to their damage to the body's DNA. They have been found in the fetal liver, lungs, heart, brain, and spleen. Fetal and placental weights are lower. There are as yet no studies on the long term damage this may cause in children.

Micro- and nanoplastics negatively affect every ecological system on this planet, from the smallest to the largest (ecosystems and ecoranges). As Machado et al comment . . .

> *It is generally accepted that the impacts of pollution on ecologically relevant endpoints (such as migratory behavior, reproduction success, and mortality) are triggered by a cascade of changes initiated at the subcellular level that propagates throughout the biological hierarchy. In this context, contaminants with broader toxicity targets can affect potentially a larger number of species and their ecological functions. As plastic particles fragment they gain novel physical and chemical properties that increase their potential interaction with organisms causing direct and indirect toxicity.*

Micro- and nanoplastics are affecting every ecological being and structure on this planet, from viruses and bacteria to fungi to plants to the tiniest soil and ocean organisms to fish to every bird and mammal species on the planet. What the long-term effects are going to be, no one knows, but they are not going to be good – and they are only going to get worse as plastic continues to accumulate, break down, and enter the air, the water, and the bodies of every living organism on this planet.

The Ecological Impacts of Pharmaceuticals and the Medical Industry

Pharmaceuticals are now major environmental pollutants, and are ubiquitous in waters and soils. Unlike other environmental pollutants pharmaceutical pollutants are not yet regulated globally . . . . But the pitfalls of pharmaceutical pollutants extend beyond acute effects to delayed effects from bioaccumulation, amplified effects from drug-drug interactions, exacerbation of drug resistance, and reduction in aquatic and terrestrial food production.

Kamba, et al

What will it mean to raise our babies on water contaminated with low levels of birth control drugs and athlete's foot remedies plus Viagra, Prozac, Valium, Claritin, Amoxicillin, Prevachol, Codeine, Flonase, Ibuprofen, Dilantin, Cozaar, Pepcid, Albuterol, Naproxen, Warfarin, Ranitidine, Diazepam, Bactroban, Lotrel, Lorazapam, Tamoxifen, Mevacor, and dozens of other potent drugs, along with hair removers, mosquito repellents, sunburn creams, musks and other fragrances? No one knows, but evidently we're going to find out, learning by doing.

Peter Montague

What is the collateral damage of the pharmacist's pipette?

Dale Pendell

Few people know (or care) that one of the world's primary and most dangerous sources of ecological destabilization is the medical industry. While it's relatively common knowledge that agricultural chemicals possess deleterious side effects, this same awareness has not extended itself to pharmaceuticals (which are often identical to agricultural chemicals). People have simply bought into the belief that the medical industry is rather benign, that pharmaceuticals are one of the great innovations of the species, and oddly enough, that such drugs are ecologically free of harm. Nothing could be further from the truth.

> *Axiom Six: Whenever a massive corporate industry that makes billions of dollars in profits every year tells you they have your best interests at heart, they are lying. In addition, they are **always** hiding things they don't want you to know. This usually involves the ways in which their business is and has always been deleterious for the health of democracies, people, the planet, and every life form on it. The only distinction to be made between such giant corporations is the degree of harm they cause.*

Here are some facts about pharmaceuticals and the medical industry that you will not have heard before . . .

- *Nearly all pharmaceuticals are made from petroleum.*

Thus, we will not be getting rid of oil. Ever. The entire medical system depends on petroleum for almost everything it needs to function, from the drugs it uses to the plastics that make its hypodermic syringes, tubing, IV bags, personal protective gear, countertops, flooring, and on and on and on. This is not the way it was fifty years ago when nearly everything it needed was reusable or recyclable. (Hypodermic syringes, for instance, were made of glass and metal and simply sterilized after use so they could be used over and over again.) Everything the medical system used was made of glass and paper and cotton and metal, most pharmaceuticals were inexpensive, and there weren't very many of them.

- *Very few pharmaceuticals are biodegradable.*

Unless exposed, for extended periods, to high heat, sunlight, or oxygen in the atmosphere, pharmaceuticals continue to be functionally active for decades or even centuries. Few drugs are exposed to those pressures. Most of the drugs we take are excreted into the water in our toilets or thrown into landfills (sometimes a small portion of the world's expired drugs are incinerated, which, of course, has its own problems).

When pharmaceuticals enter the waste stream, sooner or later they become part of the streams, rivers, lakes, or underground aquifers of this planet. Sooner or later, those thrown into landfills contaminate groundwater as well. As only one example, researchers have tracked a plume of contaminated groundwater from a landfill at Jackson Naval Air Station in Florida that has been slowly moving underground for more than forty years now. It still contains metabolically active drugs including pentobarbital, meprobamate, and phensuximide – a barbiturate, a tranquilizer and an anticonvulsant.

- *When people take pharmaceuticals, the largest portion of the ingested drugs, either in their pure form or as metabolized by-products, are excreted into the toilet, where they immediately enter the waste stream, in which they remain active indefinitely.*

Pharmaceuticals are not foods and most of the time the body does not use them similarly, that is, the majority of them are not broken down and their molecules incorporated into our bodies the way that food is. They are excreted, either unchanged or in metabolized form, meaning that as the body processes them it creates biological by-products which are then also excreted. Fifty to ninety-five percent of the drugs (depending on the drug) that people take are excreted chemically unchanged or unmetabolized into the waste stream. And they remain pharmacologically active once they are.

Nearly all medical drugs are intended to force the body to function within a certain range that researchers have decided is "normal." That is, drugs for high blood pressure force the high pressure to reduce by making the body behave. They do not cure the disease (or condition) that is causing the high blood pressure. You take the drug, the body is forcibly altered, the drug is excreted during the day, its actions slowly fall, blood pressure rises again, and you have to take the drug once more. That is why drugs have to be taken throughout the day. (What determines how often you take it is the drug's half life, that is, how quickly the drug leaves the body.)

Many people take these kinds of drugs for the rest of their lives. And the pharmaceutical industry likes it that way. They love drugs that have to be taken for a lifetime in comparison to drugs such as antibiotics that are taken short term (and actually cure a disease condition). This is one of the main reasons that the world's drug companies are getting out of the antibiotic business. (This and the fact that they have realized that bacteria will always find a way around antibiotics; antibiotic-resistant bacteria are not going to go away.)

Daily-use drugs make pharmaceutical companies money forever. And billions of people all over the planet are taking them and excreting them or their metabolites into waste streams every day of their lives – for years and years and years.

- *Water treatment facilities are unable to remove the hundreds to thousands of pharmacuticals that enter the waste stream every day.*

Waste treatment is still locked into a late nineteenth-century, early-to-mid-twentieth-century mindset. It isn't very good nor is it ecologically oriented.

> *This is a main reason that innovation in waste treatment is extremely rare – the other big factor is the western world's people's extreme discomfort with (and fear of) poop and pee and menstruation. (There is a widespread belief that humans are angelic beings sitting on top of biological sewers. As Alice Walker once put it, from my inexpert memory, "I was taught to wash down as far as possible, then up as far as possible, then to wash possible." We don't like to talk openly about possible.)*
>
> *The current fanatical use of wet wipes instead of toilet paper is to get all that nasty poop off people's asses. The wipes are then flushed down the toilet in their millions along with additional millions of tampax every day of every week of every year. These combine with fats from cooking and become the source*

> *of all those fatbergs blocking sewer systems in cities around the world.*
>
> *Individual waste treatment systems do exist that are about the size of a home heating/air conditioning unit and can be placed in any house. They treat the water so well that it is purer than tap water; it is then recycled back into the home water supply. The solids are processed into a sterile powder. Systems like these are not allowed to be used. They don't pass building codes. And I don't know if they ever will. (Drinking old pee water? No way! We are a civilized people!)*

For the most part, waste treatment remains focused on the big stuff (floaters) and some common infectious agents, not the thousands of drugs that have been created since waste treatment plants were invented.

Most people tend to think of waste treatment plants as treating household excretions and waste but at least half (and sometimes far more) of what goes into waste treatment plants now comes from industry. This includes all kinds of manufacturers such as chemical plants; hospitals (who dump massive amounts of drug-contaminated human waste, expired medications, and other wastes into the water and solid waste streams); other medical "care facilities" (physician, dentist, and veterinary offices and nursing homes flush as much as 250 million pounds of expired or excreted pharmaceuticals down the drain every year in the U.S.); mortuaries (who liquify internal organs and flush them down the drains – organs that are often highly toxic from end-of-life medical treatments – as well as massive amounts of embalming fluids and cosmetics); and pharmaceutical manufacturing and bottling plants. Very few waste treatment plants are designed to properly deal with either the chemicals that these sources dump into them or those from people's toilets. It is a case of nineteenth- and early twentieth-century technology meeting twenty-first-century waste products. And as is true of all the current infrastructure problems in the United States, no one wants to pay for upgrading the system.

As the waste streams from homes and industry flow into and through treatment plants, two things come out: liquid and sludge. Both contain significant amounts of pharmaceuticals (including those from illegal drug labs as well as drug users – an unforeseen side effect of scientists and the medical industry; they created those substances to begin with) and personal care products such as the sunscreens, lotions, perfumes, hair conditioners, and shampoo that wash off during bathing. The sludge is either put into landfills or spread on fields as fertilizer in the United States (but not for growing human food as it traditionally was before the modern, western era). In many parts of Asia sludge is still used as fertilizer, sometimes in perfect safety (as it has been for thousands of years) but with the advent of western pharmaceuticals it is often no longer safe to do so, because . . .

Most of the world's pharmaceutical production has shifted to three countries: China, India, and Pakistan. As Muhammad Saif Ur Rehman et al comment:

> *These countries have made tremendous progress in the pharmaceutical sector but most of the industrial units discharge wastewater into domestic sewage networks without any treatment. The application of untreated wastewater (industrial and domestic) and biosolids (sewage, sludge, and manure) in agriculture causes the contamination of surface water, soil, groundwater, and the entire food web with pharmaceutical compounds (PCs) and their metabolites and transformed products (TPs), and multidrug resistant microbes.*

Though people in the United States (and most western nations) generally believe that wastewater treatments plants make the liquids ecologically safe, they don't. The liquids still contain hundreds to thousands of pharmaceuticals. And all of it flows into the nation's groundwater. The water supplies of every major metropolitan area in the United States have been found to contain pharmaceuticals. When the Associated Press contacted 62 water suppliers, 34 of them reported that they do not and never have tested their water for the presence of pharmaceuticals. Those that did only tested for a few. (The machines that test need a "software library" to identify the various chemical structures; the software is very expensive. Most treatment plants can't afford it or can only afford to identify a few.) As Mompelat et al note, "Through this review, it appears that the pharmaceutical risk must be considered even in drinking water where concentrations are very low. Moreover, there is a lack of research for by-products (metabolites and transformation products), characterization, occurrence, and fate in all water types and especially in drinking water."

The problems are pervasive: pharmaceuticals are in every water source across the planet. As Maryna Strokal, a scientist at Wageningen University and Research, puts it, "In 2000, sewage was a source of pollution in about 50% of the rivers of the world. By 2010, sewage was a source of pollution in almost all rivers worldwide." (That ten-year change gives a good idea of the exponential growth of pharmaceutical use and pollution; it's only going to get worse.)

- *Excreted and waste pharmaceuticals are altering the physiology and behavior of every organism on this planet. The full range of effects is unknown, neither the pharmaceutical nor the medical industry want it to be known, hence there is little money being set aside to fund research.*

Because all Earth organisms come from common roots, human drugs affect every life form on this planet. In other words, if the drugs affect us, they have impacts on the physiology and functioning of everything else, from microbes to insects to fish to birds to mammals. Because this is rarely a focus of ecological studies, it is not yet known how severely the life forms on this planet are going to be affected, especially the long term. As researchers Christian Daughton and Thomas Ternes note . . .

> *Although most pharmaceuticals are designed to target specific metabolic pathways in humans and domestic animals, they can have numerous often unknown effects on metabolic systems of nontarget organisms, especially invertebrates. Although many nontarget organisms share certain receptors with humans, effects on nontarget organisms are usually unknown. It is important to recognize that for many drugs, their specific*

modes of action even in the target species are also unknown. For these drugs, it is impossible to predict what effects they might have on nontarget organisms.

And as Kolpin et al comment, "Surprisingly, little is known about the extent of environmental occurrence, transport, and ultimate fate of many synthetic organic chemicals after their intended use, particularly hormonally active chemicals, personal care products, and pharmaceuticals that are designed to stimulate a physiological response in humans, plants and animals."

Nevertheless, the effects that are already known are extremely frightening, pervasive, and extensive. They are also unexpected. For example, as Arnold, et al comment, "Pharmaceuticals are designed to alter physiology at low doses and so can be particularly potent contaminants. The near extinction of Asian vultures following exposure to diclofenac is the key example where exposure to a pharmaceutical caused a population-level impact on non-target wildlife."

All chemical manufacturers, including pharmaceutical companies, and most governments adhere to the Kehoe Paradigm when it comes to chemicals, which includes pharmaceuticals. Thus, the majority of all manufactured chemicals (and all pharmaceuticals) are assumed to be ecologically safe until proven otherwise. (This is why the corporations hire scientists to continually cast doubt on the harm they cause.)

Most people incorrectly believe that if the FDA determines a drug to be safe for human use it is then safe in the larger, more expansive sense of that word – that is, it is ecologically benign. But drugs are not ecologically benign. They are some of the most dangerous substances on this planet.

- *Contrary to the beliefs of the majority of medical researchers, physicians, chemists, toxicologists, reductionists, and the general population, the smaller the dose, the more physiologically and ecologically damaging the drug often becomes in the environment.*

Excreted drugs are heavily diluted by the water into which they are excreted or thrown. This means that when water (both wild and domestic) is analyzed for the presence of pharmaceuticals, the drugs are generally found to be present in parts per million (ppm), parts per billion (ppb), or parts per trillion (ppt). (As yet, no one is testing for parts per quadrillion though there is every expectation, based on what has been found so far, that they will also be physiologically active at that dilution.) Even at these incredibly tiny amounts the drugs have significant physiological impacts.

For example, Chris Metcalf, a researcher at Trent University in Ontario, Canada, detected estrone (a type of estrogen) levels in wastewater effluent up to 400 ppt and the synthetic hormone ethinylestradiol (from birth control pills) up to 14 ppt. (He found anticancer agents, psychiatric drugs, and anti-inflammatory compounds as well.) Metcalf exposed Japanese medakas (a type of fish) to concentrations typical of wastewater streams for 100 days. At concentrations of 0.1 ppt of ethinylestradiol and 10 ppt estrone the fish began to exhibit intersexual changes (showing both male and female characteristics). At 1000 ppt all the males transformed into females. That is parts per *trillion*.

These kinds of effects are not uncommon. Louis Guillette, a reproductive endocrinologist and professor at the University of Florida, spent a lifetime studying endocrine-disrupting chemicals in the environment. One area of focus was pharmaceutical estrogens and estrogen-mimics in water supplies and streams. He found that the chemicals caused reproductive problems in a wide variety of animals: panthers, birds, fish, alligators, frogs, bats, and turtles. This included, in some instances, the complete feminization of males. Androgen levels, ratios, and the amount of free testosterone in the body were all significantly altered. And the amounts needed to do this were incredibly tiny. As he noted . . .

> *We did not [test] for one part per trillion for the contaminant, as we assumed that was too low. Well, we were wrong. It ends up that everything from a hundred parts per trillion to ten parts per million are ecologically relevant . . . at these levels there is sex reversal. . . . [And the research] shows that the highest dose does not always give the greatest response. That has been a very disturbing issue for many people trying to do risk assessment in toxicology.*

There is every reason to believe that the many reproductive alterations and problems that the human species is now experiencing come in large part from pharmaceuticals and other chemicals that mimic our reproductive hormones and that we are ingesting in the water we drink. We are not exempt from the ecological realities of pharmaceuticals or any other chemical compounds that are released into the soils, air, and water of this planet. We are ecological beings on an ecological planet.

- *The amount and variety of pharmaceuticals that are entering the soils and waters of the planet are massive in scope and are increasing yearly.*

In 1999 Americans filled 2.8 billion prescriptions covering roughly 66 classes of pharmaceuticals. By 2021 that had risen to 4.55 billion prescriptions a year. By 2025 it is expected to hit 5 billion per year. These include: antidepressants, tranquilizers and psychiatric drugs; cancer (chemotherapy) drugs; painkillers; anti-inflammatories; antihypertensives; antiseptics; fungicides; anti-epileptics; bronchodilators; lipid regulators, i.e. statins; muscle relaxants; oral contraceptives; anorectics (diet medication); synthetic hormones; antibiotics; and a great many more.

The most prescribed medications in the United States are lisinopril (an ACE inhibitor for high blood pressure – 105 million prescriptions, taken daily); atorvastatin (for reducing cholesterol, preventing stroke, and reducing chance of heart attack – 105 million prescriptions, taken daily; note: many physicians want to see *every* American on statins, permanently); levothyroxine (for hypothyroidism – 102 million prescriptions, taken daily); metformin (type 2 diabetes – 79 million prescriptions, taken daily); amlodipine (for high blood pressure, chest pain, coronary artery disease – 73 million prescriptions, taken daily); metoprolol (high blood pressure and chest pain – 68 million prescriptions); omeprazole (gastroesophageal reflux disease, GERD – 59 million prescriptions, taken daily); simvastatin (a statin – 57 million prescriptions, taken daily); albuterol (inhaler for asthma, COPD, airway disease – 51 million prescriptions, used daily); gabapentin (for seizures – 46 million prescriptions, taken daily); sertraline (depression, OCD, panic attacks, PTSD, anxiety – 38 million prescriptions, taken daily); escitalopram (depression, anxiety –

26 million prescriptions, taken daily); alprazolam (panic, anxiety – 26 million prescriptions, taken daily). There are, of course, hundreds more. These are just some of the most commonly prescribed.

Until 1992, estrogens for menopause were the fourth most commonly prescribed pharmaceutical in the United States with 92 million prescriptions daily. While numbers have dropped significantly (once their long-term side effects became known) they are still present and actively disrupting the ecosystems of the planet. They have not yet biodegraded.

These numbers apply only to the United States. And while Americans take far more pharmaceuticals (on average) than people in other countries, there are billions of people around the world that do take them right along with us, every day of their lives. As Francesco Bregoli, a researcher at the IHE Delft Institute for Water Education in the Netherlands and a leader of the team that developed methods for tracking drug pollution hot spots, has said, "Technology alone will not solve the problem; we need a substantial reduction in consumption."

Reduction, however, is not going to happen. As Tim aus der Beek et al comment, "The practice of modern medicine cannot be imagined without pharmaceuticals." Or as this is better known, "Hey! We're talking about survival here!" When people are scared about their survival, they don't care about the environment. At all.

- *Pharmaceuticals in their aggregate effects are not generally additive but synergistic. That is, they combine to produce unusual and generally unknown ecological effects. And those combined effects produce impacts that are not predictable from knowledge of the individual drugs alone.*

This is a common problem in medicine itself – the prescribing of multiple medications to patients with no awareness by the physician of their synergistic effects. Every year this leads to a significant number of deaths in the people physicians treat. (According to the *Journal of the American Medical Association* about 2.2 million people are permanently disabled or hospitalized each and every year from properly prescribed pharmaceuticals, more than 100,000 die.)

The ecological impacts are far worse. Since World War II, literally trillions of pounds of pharmaceuticals of every sort (as well as a multitude of other pharmaceutical and agricultural chemicals) have been dumped into the soils and waters and air of this planet. That they produce a wide range of unexpected effects when combined is known. What those combined effects are, how extensive they are, how serious it is . . . all of that is unknown. No one is studying the synergistic effects of combined pharmaceuticals in any depth. We, and the Earth itself, are all lab animals in a vast, uncontrolled experiment for which no scientist, physician, researcher, company, or government is taking responsibility.

- *Pharmaceuticals are not a regulated pollutant in the United States or, for the most part, anywhere else. In consequence, manufacturers, hospitals, and mortuaries are exempt from the ecological impacts of their waste, most of which they intentionally put into wastewater streams.*

Because the medical industry is the source of "modern medicine" it is almost always exempt from ecological oversight. (This is a perfect example of the conflict between competing goods, human health versus planetary health.) In one West Virginia factory, for instance, owned by the generic-drug manufacturer Mylan (which is infamous for its owners raising the cost of EpiPens 400 percent), huge machines mix drug ingredients, press them into tablets, and fill capsules. As Natasha Gilbert, in her article "Dump It Down the Drain," reports, "By the end of each run, the walls, ceilings, floors, and nearly every nook and cranny of the intricate equipment were caked in powdery drug residues." The powder was in fact everywhere. As she continues, "It was standard practice, the former workers said, to then hose down some of the rooms and machines for up to eight hours and then spray them with alcohol to clear the remaining drug residues, and the wastewater would flow down a drain in the center of each room."

The pharmaceutical wastewater from that manufacturing plant flows into the local municipal treatment facility, but as is commonly true, that facility is not equipped to remove the contaminants from the wastewater stream. Researchers analyzing water downstream from the treatment plant found that, among other things, one anti-seizure medication was "90 times the amount considered safe for wildlife."

Hydrologists at the United States Geological Survey (USGS) found "substantially elevated amounts of 33 different drugs in their wastewater after testing water downstream from the factories they studied." The USGS commented that "pharmaceutical manufacturing facilities are a significant source of pharmaceutical ingredients in the environment." In fact drugs downstream from the manufacturing plants were often at levels thousands of times higher than found in rivers without such plants. (To be clear here, *all* rivers studied have been found to have pharmaceuticals in them, just as they do plastics. The levels are just smaller than the amounts found downstream from waste treatment plants.)

All those pharmaceuticals tend to bioaccumulate in aquatic insects. As one researcher noted, these insects "can have drugs at concentrations thousands of times higher in their body than in the water. They are basically small pills crawling about on the bottom of the water waiting to get eaten by fish." And the bioaccumulation just moves up the food chain, from insect to fish to what eats the fish to people.

There are thousands of drug manufacturing plants worldwide. This story is repeated in every one, across the globe. For example, at a Pfizer plant in Puerto Rico, the USGS measured fluconazole, a fungicide, at 2,000 times the levels considered safe for wildlife. (Note: the safety levels that academics come up with are guesses only. They are not actual, real-world-impact safety levels. Quite often, years into their research, they find that the safety levels were not stringent enough.)

Fluconazole, a triazole antifungal, is very similar to agricultural triazoles used on agricultural crops as well as plants that are part of, for instance, hospital landscaping. It turns out that triazoles in the environment, fluconazole or otherwise, cause resistance among fungal organisms such as *Candida auris*, which infects people and for which no known treatment exists. In one hospital outbreak, researchers found that the landscaping around the hospital was being sprayed with the same antifungal that patients were being treated with. When they had an outbreak of *Candida auris*, they closed off the hospital rooms that were infected, placed machines in the rooms that vaporized hydrogen peroxide, left them on for several days, and then retested the rooms.

Every infectious organism had been killed except for *Candida auris*. It remained unaffected.

And all these drugs that are going into the environment? They cause system-wide alterations in behavior, from microbes upward.

For instance, carbamazepine, an anti-seizure medication, is commonly found downstream from the manufacturing plants that make it. It has been found to interfere with the ability of parent fish to protect their offspring, leading "male fish to perform worse when defending their offspring from predators." As one researcher noted, "We observed higher mortality from predation . . . [the parent fish] were sluggish, so their offspring were eaten more often."

Pharmaceutical manufacturers continue to deny that any of the effects being seen are from their manufacturing plants. The causes, they say, are probably hospitals and individuals pouring their medications down their toilets.

- *The ecological impacts are pervasive and extensive.*

While the pharmaceutical companies and their minions are still denying that humans are being affected by pharmaceutical pollution (in a tobacco and climate warming kind of way) it is already common knowledge in environmental journals (though not widely reported in the media) that every other life form on this planet *is* being affected. What follows is just a rough run-through of available data (a full treatment would need a very large book in and of itself).

Benzodiazepines (anti-anxiety, insomnia, and panic disorder medications) bind to neuroreceptors in the brain and enhance the effect of a neurotransmitter called GABA. (We are not the only organism on the planet with GABA receptors in our neural system.) Fish that are exposed to this drug in the waters they live in bioaccumulate it in their bodies; levels are often six times that of the water they swim in. The drugs interfere with the normal predator surveillance behaviors of the fish as well as their social behavior with each other. The fish are less social, more active, aggressive, and bold. They are less concerned with avoiding their usual predators – which has an effect on their survivability. They eat more, and eat more quickly and aggressively as well, which is impacting the delicate food webs in which they live.

Steroid estrogens in water are now known to "correlate with widespread sexual disruption in wild fish populations." But they are not limited to impacts on fish. Environmental chemical contaminants, including estrogenic pharmaceuticals, are altering epigenetic programming in species from plants to alligators. There are "perturbations of the reproductive system including abnormal ovarian morphology, decreased robustness of sexually dimorphic gene expression with the gonad, and altered levels of circulating sex steroids." Organisms from frogs to fish to alligators to panthers are showing reproductive abnormalities. Male bass in the Potomac River, for instance, are now regularly producing eggs, not just the females. As Guillette, et al comment, "Reproductive disorders in wildlife include altered fertility, reduced viability of offspring, impaired hormone secretion or activity, and modified reproductive anatomy."

Rebecca Giggs in *The Atlantic* reports that "a platypus living in a contaminated stream in Melbourne is already likely to ingest more than half a recommended adult dose of antidepressants every day.... Amphetamines change the timing of aquatic insect development. Antidepressants impede cuttlefish's learning and memory, and cause freshwater snails to peel off rocks. Drugs that affect serotonin levels in humans cause shore crabs to exhibit 'risky behavior,' and female starlings to become less attractive to males (who in turn sing less). Dosed with Prozac, shrimp are more likely to swim toward a light source [which gets them eaten] ... and Atlantic salmon smelts exposed to benzodiazepines (such as Valium and Xanax) migrate nearly twice as quickly as their unmedicated counterparts ... arriving at the sea in an undeveloped state and before seasonal conditions are favorable" for their survival.

Benzodiazepines are extremely pervasive in the environment simply because they are some of the most commonly prescribed medications in the United States. They are often halogenated, which means that a halogen molecule is included as part of their chemical structure. This enhances their effects in the body but it also makes them far less biodegradable when they enter the waste stream. And, similar to other pharmaceutical pollutants, they have potent effects at tiny levels. As Hughes et al comment, "Antidepressants appear to pose particular risk to all taxa except bacteria with effective concentrations ranging from ug to mg L^{-1}. Invertebrates and fish show chronic toxic effects at sub mg L^{-1} levels for cardiovascular drugs and Others; fish also appear susceptible to painkillers with median effects manifesting at 40 ug L^{-1}."

During the past four decades researchers have found that many aquatic organisms, especially bottom feeders and filter feeders (e.g. shrimp, flounders, oysters), possess a special excretory system called the multixenobiotic transport system (MTS). It is composed of proteins (such as Pgp) that facilitate the removal of toxic substances from inside their cells. Because of their nature both filter feeders and bottom feeders encounter large numbers of toxins in their diet. (One of the crucial ecological actions of filter feeders, such as oysters, is to clean the Earth's waterways of toxins.) These types of aquatic dwellers depend heavily on the MTS, otherwise toxins would build up to insupportable levels in their bodies. But it is a nonspecific system; it recognizes many pesticides, drugs, and natural toxins alike as substances that need to be sequestered and removed. This has led to serious problems.

Drugs such as verapamil (a cardiac calcium ion influx inhibitor) directly bind to the receptor cite of Pgp thus limiting the effectiveness of the MTS system and its cellular pumping mechanisms. As a result toxins become more dangerous to many aquatic organisms at lower levels. Daughton and Ternes note that "Exposure to verapamil at micromolar concentrations and lower greatly increases the toxicity of a number of drugs or other xenobiotics for many aquatic organisms as the toxicant cannot be readily removed from the exposed organism." Other drugs that have been shown to inhibit the MTS include reserpine (antihypertensive), trifluoperazine (antipsychotic tranquilizer), cyclosporins (immunosuppressants), quinidine and amiodarone (anti-arrythmics), anthracyclines (noncytotoxic cytoxin analogs), and progesterone (steroid).

Selective serotonin reuptake inhibitors (SSRIs) like Prozac, Zoloft, Luvox, and Paxil have exceptionally strong impacts on aquatic organisms as well – even in tiny amounts of parts per billion. Serotonin is important in invertebrate and vertebrate nervous systems but it also plays key roles in physiologic regulatory activities in many life forms. Among shellfish serotonin regulates reproductive activities (such as spawning, egg maturation, and hatching), heartbeat rhythm, feeding, biting, swimming patterns, cilia movement, and larval metamorphosis. Among crustaceans it stimulates the release of many different neurohormones that affect such things as glucose uptake, shell color, molting, egg maturation, and levels of neuroactivity.

Some commercial shellfish farmers have long added serotonin to their crops of shellfish to stimulate spawning. Researchers, however, have found that Prozac and Luvox are the most potent such compounds ever produced, having *significant* effects at parts per billion. Extremely low doses of Prozac initiated significant spawning activity in mussels while Luvox was even stronger – dosages magnitudes smaller produced significant effects. SSRIs have also been found to significantly affect fingernail claims, mussels, fiddler crabs, crayfish, snails, squids, and lobsters with wide-ranging effects at extremely low doses. Pharmaceutical SSRIs are some of the most widely dispensed drugs in the industrialized nations. But they are not the only drugs that have been found to affect crustacean reproduction.

Fenfluramine, a sympathomimetic amine, once popularly prescribed as a diet drug (removed from the market in 1998 because of heart valve damage in patients), has also shown strong reproductive system activity in crustaceans at low doses: it triggers ovary-stimulating hormones in crayfish and gonad-stimulating hormones in male fiddler crabs. And retinoids, prescribed in large quantities for such things as acne (Accutane), cancers such as leukemia (Vesanoid), and wrinkles (Retin-A or tretinoin, an anti-aging prescription and one of the top 200 most widely prescribed drugs in the U.S.), have been shown to have profound effects on amphibian embryonic systems. Constant exposure can produce deformities in the offspring of frogs and other amphibians.

- *But the most dangerous of all pharmaceuticals are the antibiotics that are being released into the environment; they are pervasive and disturbing ecological systems that are foundational to the entire functioning of this planet. They are also stimulating the emergence of antibiotic resistant organisms at an exponential rate. As a number of researchers have said, The Age of Antibiotics is over. We now face the rise of pathogenic organisms more terrible than any known before.*

Most people are now aware that all of us possess a microbiome in our intestines, that is, we have a microbial community inside us upon which our health depends. When it is healthy, the bacterial organisms in our gastrointestinal tract help us digest our food, provide substances that we need to be healthy, keep our immune system strong, and keep our organ systems functioning well, including our brains. (We also have microbiomes in our lungs and on our skin that provide similar benefits.)

The bacteria that make up our microbiome have been part of human bodies since human beings have been; they have been transferred from mother to child from the beginning of species time. Without them we would not be healthy, we would not even be alive. They are essential to our lives and existence. They are in fact part of us just as we are part of them. And the truth is that nearly all bacteria are friendly. Of the millions of different kinds of bacteria only a few are pathogenic to human beings.

However, when people take antibiotics their healthy microbiome is disturbed. More plainly, large numbers of friendly bacteria are killed and the entire community and its functioning is damaged. This is why antibiotics cause so many problems in the gut: nausea, indigestion, bloating, vomiting, severe cramping, diarrhea, and blood or mucus in the stool. Usually, when the antibiotics are stopped, the intestinal microbiome, after a few weeks, rebuilds itself. It recovers.

But if people keep taking the antibiotics, the microbiome cannot recover and more side effects will occur. This includes such things as fever and chills, out-of-control infections as pathogenic organisms take advantage of the loss of the protective function a healthy microbiome provides, generalized pain throughout the body, light sensitivity, rapid heartbeat, skin rash, dizziness, swelling, wheezing, coughing, difficulty breathing, low blood pressure, fainting, seizures, and the emergence of long-term, chronic diseases such as diabetes, obesity, inflammatory bowel disease, asthma, rheumatoid arthritis, depression, and alteration of mental functioning, which includes things like brain fog, forgetfulness, trouble concentrating, depression, depersonalization, suicidal tendencies, and a large variety of other disturbed mental states.

What most people do not know is that Earth itself has a microbiome. It extends from miles below the planet's surface to miles upward into the atmosphere. It covers the entire surface of the Earth as well and every organism on it, including, like us, their interiors. Similarly to people, the Earth depends on its microbiome for healthy functioning as does every complex life form on this planet. When those microbiomes are disturbed by antibiotics, the same kinds of disease and malfunction begins to occur throughout the Earth's ecosystems and its life forms. And over the past seventy-five years every microbiome, including that of the planet, has been disturbed. Significantly so.

Human beings discovered antibiotics prior to World War II but they did not become part of standard-practice medicine until 1946. In 1942 the world's entire supply of penicillin (the first antibiotic) was 64 pounds. By 1949, 156,000 pounds a year of penicillin and a new antibiotic, streptomycin (from soil fungi), were being produced. By 1999, in the United States alone, this figure had grown to 40 million pounds a year. By 2009 it was 60 million pounds a year and, of course, millions of pounds more in countries around the world. This is *every year, year in and year out*. And these numbers are increasing all the time. Similarly to other pharmaceuticals, antibiotics are not easily biodegradable.

In an extremely short period of geologic time the Earth has been saturated with several *billion* pounds of non-biodegradable, often biologically unique pharmaceuticals designed to kill bacteria. Most antibiotics (literally meaning "against life") are what are called "broad-spectrum," meaning they do not discriminate in their activity but kill broad groups of diverse bacteria whenever they are used. The worldwide environmental dumping, over the past four decades, of such huge quantities of synthetic antibiotics has initiated the most pervasive impacts on the Earth's bacterial underpinnings since oxygen-generating bacteria supplanted methanogens 2.5 billion years ago. As bacterial researcher Stuart Levy comments, "It has stimulated evolutionary changes that are unparalleled in recorded biologic history." In other words, the entire microbiome of the planet and every life form on it is experiencing a severe and unremitting disturbance. The antibacterial disturbance that "modern medicine" has caused to Earth's microbiome is one of the most dangerous technological impacts that unrestrained corporate industry has created. It is far more serious than climate change for it is a direct threat to every life form on this planet, including Earth itself. The tip of this iceberg, the one that most people have heard about, is the rise of antibiotic-resistant bacteria.

The story that is most commonly told about the rise of resistant organisms is terribly oversimplified and in many respects inaccurate. What you have probably heard is that when we take an antibiotic, the antibiotic kills off the susceptible bacteria but there are always a few that are resistant for one reason or another and these survive to have offspring and thus we have the rise of resistant organisms. (And then of course, there is the inevitable mutation that happens every so often.) These stories are not accurate; in fact they come from a deep misunderstanding of what bacteria are and what they can do (it is in fact a remnant of late nineteenth- and early twentieth-century beliefs). What is really happening is far more complex and dangerous to human beings.

In response to the billion of tons of antibiotics flooding the ecosystems of the planet bacteria have generated highly sophisticated alterations in their physiology and behavior. They have literally begun rearranging their genomes in order to make their bodies resistant to the antibiotics. As their genomes shift, bacterial physical structures alter, sometimes considerably. They are, literally, remaking themselves and their communities so they can better respond to this threat to their existence. And it is happening all over the planet, to every bacterial organism there is. In consequence, the entire microbiome of the planet, relatively stable for 2.5 billion years, is altering itself in ways that are shifting the entire microbiome functioning of the planet. No one knows what our world is going to look like as it does. But the last time this happened, 2.5 billion years ago, the very nature of life on this planet changed and it never went back to the way it had been before.

Bacteria, as soon as they encounter an antibiotic that can affect them, however minutely, begin actively generating possible solutions to it. The variety and number of the solutions they can generate are immense, from inactivating the part of the bacterial cell that the antibiotic is designed to destroy, to pumping the antibiotic out of their cells just as fast as it comes in, to altering the nature of their cellular wall to make it more impervious, even to using the antibiotic for food. And these solutions? They are passed on to their descendants. In essence, it's the passing on of acquired characteristics, something Lamarck insisted was possible and that the neo-darwinians have ridiculed ever since.

Ironically enough, it was Alexander Fleming, the discoverer of penicillin, who first warned of bacterial resistance. He noted as early as 1929 in the *British Journal of Experimental Pathology* that numerous bacteria were already resistant to the drug he had discovered and by 1945 he warned in a *New York Times* interview that improper use of penicillin would inevitably lead to the development of resistant bacteria.

At the time of his interview just 14 percent of *Staphylococcus aureus* bacteria were resistant to penicillin – by 1953, as the use of penicillin became widespread, 64 to 80 percent of the bacteria had become resistant and resistance to tetracycline and erythromycin were also being reported. (In 1995 an incredible 95 percent of staph organisms were resistant to penicillin.) By 1960, resistant staph had become the most common source of hospital-acquired infections worldwide. (This is known as an exponential growth curve.)

So, physicians began to use methicillin, a β-lactam antibiotic that they found to be effective against penicillin-resistant strains. Methicillin-resistant staph (MRSA) emerged within a year. The first severe outbreak in hospitals occurred in the U.S. in 1968 – only eight years later. Eventually MRSA strains resistant to all clinically available antibiotics except the glycopeptides (vancomycin and teicoplanin) emerged. But by 1999, fifty-four years after the commercial production of antibiotics, the first staph strain resistant to all clinical antibiotics had infected its first three people.

This rate of resistance development was supposed to be impossible. Evolutionary biologists had insisted that evolution in bacteria (as in all species) could only come from spontaneous, usable mutations that occur with an extremely low frequency (one out of every 10 million to one in 10 billion mutations), each generation. That bacteria could generate significant resistance to antibiotics in only thirty-five years was considered impossible. That the human species could be facing the end of antibiotics only sixty years after their introduction was ludicrous.

Bacteria are the oldest forms of life on this planet and they have learned, during that time span, how to respond to threats to their wellbeing. Among those threats are the thousands if not millions of antibacterial substances that have existed for as long as life itself has. The world is, in fact, filled with antibacterial substances, most produced by other bacteria, fungi, and plants. As Steven Projan of Wyeth Research puts it, bacteria "are the oldest of living organisms and thus have been subject to three billion years of evolution in harsh environments and therefore have been selected to withstand chemical assault." And our antibiotics? Most of them are actually just slight alterations of antibacterial substances that are already common throughout the natural world – substances that bacteria have long been aware of. (But, of course, those substances existed at much lower concentrations until we came along.)

Once a bacterium develops a method for countering an antibiotic, it systematically begins to pass the knowledge on to other bacteria – not just its offspring – at an extremely rapid rate. Under the pressure of antibiotics, bacteria are interacting with as many other forms and numbers of bacteria as they can. In fact, bacteria are communicating across bacterial species, genus, and family lines, something they were never known to do before the advent of commercial antibiotics. And the first thing they share? Well, it's resistance information.

Bacteria can share resistance information directly or simply extrude it from their cells, allowing it to be picked up later by roving bacteria. They often experiment, combining resistance information from multiple sources in unique ways that increase resistance, generate new resistance pathways, or even stimulate resistance forms that are not yet necessary. Even bacteria in hibernating or moribund states will share whatever information on resistance they have with any bacteria that encounter them.

Bacteria experiment and innovate. Their main laboratories for developing resistance are places where ill people congregate: hospitals, nursing homes, prisons, schools. The massive use of antibacterial substances in hospitals and nursing homes allow multiple species of bacteria exposure to them, plenty of time to innovate, and the easy transference of resistance information to one another. When new bacteria take up encoded information on resistance, they weave it into their own DNA and this acquired resistance becomes a genetic trait that will be passed on to their descendants forever. As Earth systems researchers Williams and Lenton comment, "Microbe transfer between local populations carries genetic information that changes species composition and thus alters the nature of each community's interaction with its local environment." And those altered interactions? They are occurring worldwide and no one knows what it will mean to life on this planet in the long run.

Bacteria are not competing with each other for resources, as standard evolutionary theory predicted, but rather promiscuously cooperating in the sharing of survival information. They are responding as a whole to the threat to their existence. Anaerobic and aerobic bacteria, Gram-positive and Gram-negative, spirochetes and plasmodial parasites, every kind of bacteria there is, all are exchanging resistance information. Something that, prior to antibiotic usage, was never known to occur.

Bacteria are acting in concert so well in response to the human "war on disease" that it has led Levy to remark, "One begins to see bacteria, not as individual species, but as a vast array of interacting constituents of an integrated microbial world." Former FDA commissioner Donald Kennedy echoes this when he states that "The evidence indicates that enteric microorganisms in animals and man, their R plasmids, and human pathogens form a linked ecosystem of their own in which action at any one point can affect every other." Or as Lynn Margulis once put it, "Bacteria are not really individuals so much as part of a single global superorganism."

Bacteria are, in fact, responding socially, as a community. As writer Valerie Brown notes: "In a series of recent findings, researchers describe bacteria that communicate in sophisticated ways, take concerted action, influence human physiology, alter human thinking and work together to bioengineer the environment."

Worryingly for the medical establishment, bacteria are also generating resistance to antibiotics researchers haven't even thought of yet. For example, after placing a single bacterial species in a nutrient solution containing sub-lethal doses of a newly developed and rare antibiotic, researchers found that within a short period of time the bacteria developed resistance to that antibiotic *and* to twelve other antibiotics that they had never before encountered – some of which were structurally dissimilar to the first. Stuart Levy observes that "it's almost as if bacteria strategically anticipate the confrontation of other drugs when they resist one."

There are billions, perhaps trillions, of different kinds of bacteria on this planet. All of them are ecologically relevant. All are important to the functioning of this planet and its life forms. Very, very few of them are dangerous to us. But they are not taking the corporate creation and environmental release of antibiotics lightly. And this has serious implications for the human species.

Human death rates from resistant organisms are rising exponentially. While the CDC's website insists that only around 23,000 resistant infectious deaths occur every year in the U.S., researchers Burham et al (2019) estimate that the true figures are at least 7-fold higher, or 162,044 deaths per year. (Worldwide, it is, at minimum, several million each year, projected to reach 10 million a year by 2050 – and these are *very* conservative estimates.) By this analysis infections from resistant bacteria are now the third leading cause of death in the United States. And it is only going to get worse. As Mark Lappe has said, "The period once euphemistically called the Age of Miracle Drugs is dead." Or as David Livermore, MD, of the Antibiotic Resistance Monitoring and Reference Laboratory in London, England, says, "It is naive to think we can win." We now face the emergence of pathogenic, pandemic organisms more terrible than any our species has known before. And there is no escape, no safe harbor, for the bacteria; the entire planet is massively polluted with antibiotics.

As Natasha Gilbert reveals in her article for *The Guardian* ("World's Rivers Awash with Dangerous Levels of Antibiotics"), "Hundreds of sites in rivers around the world from the Thames to the Tigris are awash with dangerously high levels of antibiotics, the largest global study on the subject has found. Antibiotic pollution is one of the key routes by which bacteria are able to develop resistance to the life-saving medicines, rendering them ineffective for human use."

The truth is that bacterial resistance is growing at an exponential rate. Bacterial researchers around the world are quite clear that we are very close to the point when antibiotics are going to fail entirely. Once they do even simple surgeries will become dangerous, infectious pandemics will arise (such as ones from *Candida auris,* which has no known treatment), "modern medicine" will collapse, for its entire success rests on the use of antimicrobials. And once resistance begins, it spreads everywhere and it spreads fast.

The bacterialiologist John Prescott comments, "There is essentially no gene in any bacterium that cannot be moved to another bacterium." Create superbugs in a waste stream and their resistance genes are going to move, create them in farm animals and they are going to move, create them in hospitals and they are going to move. And they do move. Hospitals (and other treatment locations from medical offices to nursing homes), agribusiness farms, and pharmaceutical manufacturing plants are creating resistance genes faster than researchers can keep up with them. "Medical professionals around the world," as Sasha Chapman comments, "warn of a post-antibiotic era, when bacteria will be resistant to all the drugs we can throw at them. The prospect is scary enough to be called a 'crisis' (by the WHO), a 'nightmare' (by the CDC), and a 'catastrophic threat' (by UK chief medical officer Sally Davies)."

While the rise of resistant organisms and the repercussions we face in and of themselves, are terribly frightening the thing to keep in mind is that if bacteria had not developed resistance, *all life on this planet would already have died.* Every form of life here depends on the bacterial microbiome of this planet and the microbiomes inside and on their bodies. Because scientists, corporations, and physicians insisted that bacteria were unintelligent and not highly adaptable, they believed they could create antibiotics and spread them around the world without consequence. The only outcome would be the end of infectious disease. But in doing so they created one of the greatest threats they could have created to all life on this planet. *It is not all about us.* We are not alone here.

Every time you take a pharmaceutical for your health, to treat disease, to extend your life span or to save it, remember: you are affecting every other life form on this planet and every ecosystem that exists – and that includes us, the human species. And what's more, there are some 8 billion people doing that right along with you every day of the week, every week of the month, every month of the year, year after year after year. Our medical system (and your health care) is not exempt from the ecological realities of this planet.

The medical system is in fact one of the most dangerous environmental polluters on the planet, the least known but one of the most powerful, and one that will fight endlessly to prevent its regulation. The medical system and its pharmaceuticals are undermining the entire ecological functioning of the planet. The only way to stop its effects is the immediate, significant reduction, by at least 90 percent, of the use of pharmaceuticals worldwide. Which is not, of course, going to happen.

Pharmaceuticals should be understood as what they are: extremely dangerous ecological poisons and systemic disruptors. Our world civilization cannot and will not survive their ecological impacts.

How Civilizations Fall

Civilizations, like people, have life spans. They are born, go through adolescence, grow strong, enter middle age and, inevitably, sooner or later, old age. It is then that their decline begins in earnest. They become senile, collapse, and die. This process of decline and fall always involves the same grouping of factors. There are six of them, each of which is synergistic with the others. The sixth is the most serious when it comes to how deep and severe the collapse will be.

The factors are:

1) *Climate change*
Civilizations depend upon a predictable climate. It allows them to forecast future group behaviors and the expected outcomes of those behaviors. As one simple example: planting x amount of a crop leads to y yield thus allowing the amount of food that is needed yearly to be efficiently planned for. Maintaining a civilization involves a continual analysis of food needs and a great many other factors, each of which is essential to its continuance. Predictions are then made to meet those future needs.

Foundational to every one of these predictions is a stable Earth climate and environment.

All civilizations assume that future climate will match past climate, though, of course, with some variance around the mean. They can survive a few unusual years but if the climate significantly changes things begin to fall apart. In other words, a fifty-year drought makes even the most elegant predictions useless.

As climate instability increases it becomes much harder to make successful future predictions. This affects every aspect of culture, from what kinds of houses to build, to heating demands, to water supplies, to clothing, to trade . . . and so on, endlessly. Large civilizations are not very adaptable to the need for rapid change, especially that of climate. After a certain point, they can't keep up and the collapse begins in earnest.

Another way to say this is that our world civilization has adapted to a very specific and unique Earth climate that has been relatively stable for the past 12,000 years. That climate is, in fact, what allowed what we think of as civilization *to* develop. Everything people know and believe and think is oriented around it. Because of the continual technological and corporate damage to Earth systems that long-term climate is destabilizing at increasing rates every year. Every ecological system is failing, no matter how large or small it is. This has happened before to civilizations in specific regions but never to the entire planet, to all ecoregions simultaneously. *As a species*, we are in uncharted territory.

This first dynamic is now fully in play.

2) *Environmental degradation*

Civilizations always degrade local ecology through the same grouping of behaviors. At its core this entails damaging, then exhausting, and ultimately destroying ecological health through unrestrained resource extraction to meet the needs of too many people in a limited location. (Earth now has too many people in a limited location: this planet.) Lack of ecological robustness and health always underlies structural failure. Loss of soil health leads to poorer crops. Increasing use of fertilizers contaminates waterways. Drought leads to insufficient water for people, crops, domestic animals, and industry. Worsening air contaminants lead to ill health and stress in ecological systems. And so on and on and on. Ultimately the civilization, that is, the virtual reality that rests on top of the real world, can no longer be supported because the source of its raw materials is depleted and broken. It can't replenish itself. Once the real world ceases to be robust, the virtual world that draws upon it for sustenance has no other option but to collapse. (It is an ecological consequence that is built in, it takes pressure off Earth systems so they can regenerate.) Our ecological systems are no longer robust due to unrestrained corporate extraction.

This second dynamic is now fully in play.

3) *Increasing complexity of government bureaucracy, infrastructure, technology*

As the complexity of a civilization increases there is a corresponding need for larger and larger bureaucracies to run the government, technology, and infrastructure that keep the wheels turning. The larger and more complex the bureaucracy is, the more unwieldy it becomes and the more it begins to serve itself rather than the civilization that created it. It gets bogged down in its own complexity, internecine struggles, and an overabundance of rules and regulations. It can no longer keep the civilization working smoothly. At a certain point it is unable to respond effectively to shocks to the system. A series of shocks occurs, the bureaucracy responds with increasing ineptness – a final external stressor hits and it fails.

This is now fully in play.

4) *External shocks (pandemics, war, famine, invasion, environmental disasters such as wildfires, volcanoes, tsunamis, drought, and so on)*

Once environmental degradation, climate problems, and bureaucratic complexity reach a certain point, the system becomes extremely unstable and very vulnerable. External shocks are met with increasingly inept responses and growing social unrest. Over time, due to systemic instability in the ecological system, external shocks occur more rapidly, the bureaucracy responds ever more poorly, and the system fails ever more seriously.

In the United States, we are already in this cycle, specifically we are experiencing what are called Black Swan events, which, when they reach a certain frequency of occurrence, is also known as Gambler's Ruin.

Since 2001, the U.S. has experienced 9/11, the unwinnable and inept invasions of Iraq and Afghanistan, the condoned use of torture during interrogations, the abandonment of the Geneva Conventions, Hurricane Katrina and the failure of the levees in New Orleans, the mortgage collapse, a worldwide financial panic and long-term depression, a severe megadrought in the west accompanied by yearly heat waves and massive wildfires, massive and ongoing refugee/immigrant crises, the election of Trump, extreme political and social polarity not seen at these levels since the decade prior to the American Civil War, the Covid-19 pandemic, widespread protests and riots in response to the police killings of African-Americans, massive financial inequality at levels not seen since the Gilded Age, the growing insistence by various large subgroups of Americans that the Constitution of the United States is an outdated document (and similar processes in many countries around the globe), continual attempts to overturn the 2020 presidential election, and the assault on the Capitol. None of these were or are being handled well. Every one of these events has been surrounded by misinformation and spin.

These events have exposed the dysfunctional response of the American bureaucracy and culture in the face of systemic shocks. The system has failed (and is failing) and the people know it. (It took five days for the – purportedly – most powerful country in the world to get water to the people sequestered in the Superdome in New Orleans after Hurricane Katrina. There were no toilets, no toilet paper, no menstrual pads, no blankets, no food, no water. Those that died remained among the living; they were stacked against the walls, there was no place else to put the bodies.)

Over time, as the cycle continues, the bureaucracy responds ever more poorly. Social unrest due to loss of faith in the system because of its inept responses continues and worsens. Splinter groups insisting that they best know how to respond to the people's needs and return the country to normalcy will emerge with ever greater frequency. And the splinter groups will, as they always do, turn on each other with ever more vehemence.

New shocks will occur over ever shorter time spans, putting more stress on the system. And those new shocks will be unexpected and unpredictable. The cycle will continue and escalate.

This factor is now fully in play.

5) *Randomness/bad luck (nonlinear fluctuations in complex systems, Gambler's Ruin)*

Sometimes bad luck happens – all human beings, sooner or later, realize this. Bad luck is built into the system, into life. Bending over and your eyeglasses falling into an outhouse pit is bad luck. Your car being stolen is bad luck. A good candidate losing an election to a bad candidate because of hanging chads is bad luck. A broken gauge in a crucial section of a nuclear power plant just as the plant begins to break down is bad luck. An unpredicted perfect storm is bad luck for a boat crew. Bad stuff happens but sometimes . . . it happens in cycles. A series of external shocks one right after another (as has been happening in the U.S.) is one particular form of bad luck. It has a name: *Gambler's Ruin*.

Bad luck, as all gamblers and people past a certain age know, tends to run in cycles – that is, everything begins to go wrong. Then that bad luck escalates in its frequency. Nothing you do is able to stop it. For gamblers, this is the point where the cards turn cold but the player keeps on playing rather than walking away. Professional gamblers have learned, the hard way, to walk away; non-professional gamblers keep playing until they lose everything. It is the all too human tendency to keep on doing the same thing but expecting a different outcome.

The overriding (erroneous) belief, with both individuals and in bureaucracies, is that the system you are using *will* eventually work, you just haven't been doing it forcefully or expertly enough. So, the decision is made to keep doing it, only more so and harder.

People who study nonlinear systems are the ones who named this process Gambler's Ruin. It's because they noticed that while statistics says flipping a coin will, over time, come up heads and tails an equal number of times, in the real world it doesn't work like that. Heads may come up 20 times in a row, tails 3 times, then heads 15. (And upsettingly, the coin may sometimes, very rarely, remain on its edge, neither heads nor tails.) Life is not statistical; it's not linear, it is not predictable. It is in fact nonlinear and all living systems, including governments, people, and Earth itself, are nonlinear in their behavior and functioning. This means that small factors can lead to large, unpredictable outcomes in the real world. So, when a systemic instability occurs in a nonlinear system, eventually, integral feedback loops begin reinforcing a pattern of collapse rather than healthy growth. This is what happens when tipping points are reached in natural systems. In common parlance, it's when life goes to shit and nothing anyone does seems able to stop it. In fact everything that is done to halt the process only increases the instability. The only way out of this is to stop doing what is no longer working and do something else. Very few people (or civilizations) can or will do this until the collapse is so bad that they are forced to give up the old patterns and stop using them. This is why real change often occurs only when stratified systems collapse.

This factor is now fully in play.

6) *Position on the rungless ladder*
Think of civilization as a kind of ladder. (Yes, this is a metaphor.) As the culture climbs upward (what many people call progress) the lower rungs (previous technological states) fall away. Some civilizations fall when only a few rungs have been passed. As with a ladder, a fall from a height of just a few rungs is easily survivable. But the higher the civilization climbs, the longer the fall will be. Eventually, once the culture reaches a sufficient height, any fall will be fatal.

The distance from the ground is defined as how far the people in a culture are from a hunter-gatherer/agrarian/low-technology lifestyle. The farther up the ladder the culture goes, the less knowledge its people, in aggregate, still have of hunter-gatherer/agrarian/low-technology lifestyles. There are always some people who do have the old skill base, of course, but as the position on the ladder increases in distance their percentage relationship to the aggregate becomes so small that their knowledge base is unable to counteract the degree of ignorance among the larger population.

In other words, with the collapse of a technological culture, knowing how to use a television remote will be a useless skill. Most people in our era, especially in the West, have adapted to a highly technological social structure. They know how to navigate it. They no longer need to depend on themselves to meet basic survival needs (in the older sense of that phrase); the culture does that for them. All they have to do to survive is to fit into the existing structure and follow its rules. They will then be sufficiently provided for. (That is, go to school, get a job, make money, buy what you need, pay the bills.) But once the structure collapses, the knowledge of how to fit into that system is useless. Instead of being a survival advantage it becomes instead a survival disadvantage.

The knowledge of how to meet low-tech survival needs is absent, that is: building shelter, gathering and growing food, making clothing, building fires, surviving the winter, interfacing with wild ecosystems, self-defense, craft and tool making, and so on and on. In consequence, the majority of the population is unable to adapt to civilizational loss. In other words, the farther the distance, the greater the fall, and the less able the group is to survive the collapse of the structure that has been holding it up.

In the past, most civilizational collapse was confined to relatively small regions; it was a temporary setback and people could more easily return to agrarian or hunter/gatherer lifestyles simply because their position on the ladder was not very high *and* they were surrounded by people who did have a deep knowledge base of agrarian/hunter-gatherer lifestyles. But in our time, the collapse will not be confined to small regions. It will be worldwide. And to make matters worse, the underlying ecological fabric to which people in the past returned has been severely damaged. There is not a healthy natural world for them to build a new life around. It will be the collapse of the Roman empire but on a worldwide scale.

This is what we now face. And I think a lot of people know it even if they are afraid to say it out loud.

> *Now . . . take a deep breath and ask yourself how you feel – in your self, in your body, in the center of who you are. What is your breathing like? Quick and shallow, high in the chest? Take another deep breath, relax, and understand that this part of the book . . . it was designed to create exactly this effect, to put you in your head, to dissociate you from your body and your*

feelings. It's the way this information is usually presented; it has a particular kind of impact, doesn't it? A particular kind of feel to it. It's not very helpful to solving the problems we face, even if it does reveal the diagnosis more easily to the eye. The important thing, however, is what you do with the information and how you experience it.

That is, rather than letting this information alter your physiology and emotional state in the way it often does, give up the dissociated state instead – give up the physical, emotional, and mental shutdown – and respond instead as a living, breathing human being. One who has feelings, far more feelings and emotions than fear. This is the only way get to what is on the other side of grief.

Because what is true is . . .

CHAPTER FOUR-THE-SECOND-PART

Inevitability and Descent

The most difficult existential dilemma we face is to at once acknowledge the bleakness before us...

Chris Hedges

The mainstream communication strategy for the last decades has been positivity and spreading inspiration to motivate people to act. Like: "Things are bad, but we can change. Just switch your light bulb." You always had to be positive, even though it was false hope. We still need to communicate positive things, but above that we need to communicate reality. In order to be able to change things we need to understand where we are at. We can't spread false hope. That's practically not a very wise thing to do. Also, it's morally wrong that people are building on false hope. So I've tried to communicate the climate crisis as it is.

Greta Thunberg

Premature reassurance and pressure to accept a loss just short-circuit the grieving and recovery process.

Phyllis Windle

The diagnosis is in fact terminal. And nothing anyone does is going to change that.

To be clear our planet is **not** dying – nor is our species. What *is* dying is our civilization. What *is* dying is our way of life. What *is* dying is the ecological form that Earth has had for the past twelve thousand years, the habitat that allowed human civilization and our many ways of being to develop and cover the planet. ***And what is true, but no one will say out loud,*** is that the only way to stop what is happening is to stop all human use and chemical modification of hydrocarbons completely totally and utterly, today. And that means right now. ***This minute.*** Not tomorrow, not by 2030 or 2100 or some other future date. ***Now.*** It is the only way the ecosystems of our planet can possibly recover. The splinter has to be removed for the body to heal.

But this will, of course, never happen because not only is there too much money involved, it would mean the end of all pharmaceuticals and all plastics (and plastics are now an integral part of nearly everything that our civilization is dependent upon including the medical industry) and gasoline and heating oil and all mass-produced electricity and computers and corporate agriculture (from which most of our food comes) and interstate and intercountry transport of goods and the building of most houses and all concrete and glass and steel, too, and the end of all our industrially created machines as well, and so on and on and on and on. And as I heard a young woman at a party say Once Upon A Time (in the midst of a terrible divorce), "Hey, all this love and light stuff is fine but we're talking about survival here." And so, later, she, in her fear, abandoned the beliefs she'd insisted were foundational to her, which is what most people do when their survival is at stake anyway. So, no, the patient is not going to give up smoking nor are they going to remove the splinter. The chemical alteration of hydrocarbons will not stop, that is, until Earth itself makes us stop (which it will even if every person on the planet drives an electric car – our problems are more complicated than that superficial solution and besides what about all those electric car batteries when they wear out, they don't just disappear you know, and what about the third-world mining that supplies the lithium for them – things are not as simple as they appear . . . ever). And when Earth does finally make us stop the modulation of hydrocarbons (one way or another) our civilization will fall. For the diagnosis is, and has been for a very long time, terminal. And no, there is nothing that can be done about it. We passed that possibility a long time ago.

Now this is the point where you throw this book against the wall. Or stop reading. Or become angry, enraged even. Or get so depressed you put the book down and begin to drink copious amounts of alcohol. (I have found Glenmorangie, a fine single malt scotch, to be the perfect choice.) Or maybe it's the point where you are certain that I have my head up my ass in some kind of new age yogic cranial-sacral inversion process that only the really naive who believe in crystal healing can do. And for sure you will begin to argue against what I have said and start to use all the rationalizations that have become so common in the press and in books by environmentalists and techno-utopianists for so long that everyone just sort of thinks them automatically now. Like the one that says if we accept that the diagnosis is terminal all that will be left is apathy (*everybody will just stop trying to do anything!!!!!*), which is not actually true and don't these people know anything about terminal diagnoses and how people in the real world actually deal with them? Or all those techno-solutions will pop into your mind, the ones that are just over the horizon and that everybody but you, Stephen, knows will stop what is happening. Like wooden skyscrapers and biofuels and wind farms and **science!**, and . . . (Just out of curiosity: do you really know where the thoughts you are thinking right now come from?) And I suspect as well you might feel hopeless and afraid at just the touch of the possibility that the diagnosis is accurate *and that the truth is there really isn't anything any of us can do to stop what is happening now or what is going to happen even more so not too long from now.*

Maybe you will do and feel all of these things. Over the years, I did and I have. Some days I still do. For I am a person, too, not that different from anyone really. It took a long time, nearly a half century in fact, for me to fully accept the diagnosis and give up false hopes, to realize that changing light bulbs or stopping single-use plastics or driving electric cars wasn't going to do diddly-poo. It took so long, in large part, because when I first got involved in all this in the ancient days of fifty years ago it was a lot easier to ignore, well, a lot of things actually, things that are impossible to ignore any longer. Because fifty years ago, the world really was a different place. The environment wasn't as damaged, the forests and the planet's ecosystems were more intact, world population was billions fewer, extraction technology was less developed, and chemical modulation of hydrocarbons was in its infancy. There were, in fact, almost no plastics at all.

Nevertheless, whether or not I was willing to fully accept the diagnosis a half century ago, I knew the trouble we were facing ever since I read *The Limits to Growth*. Even then I had a very good grasp of exponential growth curves and knew what happens when they are ignored ... whether it is the growth of antibiotic resistant bacteria, of the human population, or of the unrestrained extraction of "resources" from the ecological "capital" (rather than the ecological "interest") of the planet. But I was young (it *was* fifty years ago after all) and I'd little experience then with the tragedy of the human condition ... or the lessons inside mythic communications such as Shakespeare's *Macbeth* or the really old greek stories like the one about Icarus. What is really true is that people aren't actually rational (they are *rationalizing* if you want to get technical about it), and while they often know what is going to happen to them if they don't stop what they are doing, they continue to do it anyway simply because of the nature and limitations of their character (which is what Shakespeare was talking about and how did one guy get so smart anyway?). And besides I believed that somehow we would find it in us as a species to do what was necessary to change our course before it was too late. Which is actually a kind of false hope when you think about it from any sort of an I've-lived-a-long-time-now perspective, that is if you know anything about human history or the peloponnesian war or stuff like that.

During my youth, I watched the birth and growth of the environmental movement and heard and read all the reports and suggestions and fantasies about what we can do to stop what is happening. (And maybe we really can put a solar shield around the planet to stop global warming and perhaps we really can shoot all our garbage into the sun and learn how to only eat seaweed *and* love it!) I believed. For so many reasons, I believed.

Yet as the years, then decades, passed few of the things that were suggested were done and those that were done didn't stop what was happening even if they might have slowed things down a little bit because none of them included stopping all chemical modification of hydrocarbons and the wide dispersal of those modified compounds in all their various forms. None of them reduced human population or the pressures that the increased population was putting on the planet's ecological structures. None of them reduced corporate monopoly, power, or control, quite the opposite in fact. And so, things went on as they had been only more so because in that half century world population went from 3.7 billion to 7.8 billion while the american population increased by 100 million and they all bought cars and houses and stuff and more forests were cut down for reasons that I am not sure were thought out at all well, and more and more modified hydrocarbons were created by corporations and technologists for all those more and more billions of people, including plastics, which were only present in very tiny amounts in 1970, and they all flowed into the body of this nonlinear, self-organized Earth, destabilizing the organism, this planet, upon which our civilization and our lives, and the lives of everything that is, depend. And eventually we reached the tipping points that were obvious a half century ago (though people keep saying we haven't and we can just go on this way probably forever and Earth really can support 100 billion people, just you wait and see) and now there is no going back to that better time and those earlier days when we really might have been able to do something to avoid where we are now. *The diagnosis is in fact terminal.*

And pretty much every person that is, really and for sure, doesn't want to know this or accept it and all the people in the environmental movement and in government and in science and technology and in economics and on and on and on are going to keep saying it isn't true (out loud anyway because they don't want to be alarmist or accused of overstating things or be too gloomy which will make people too depressed and apathetic and besides everybody will get mad if they find out they have been bullshitted all this time and that, yes, things are going to get pretty bad before too long, if they aren't already bad where you live, and you don't still live in California, do you?) and besides if we just *[insert fantastical solutions you have read or heard about here]* then we can turn things around and just keep on doing what we have been doing. Forever and ever and ever. Which, of course reminds me of a story my therapist told me long ago when he was trying to get through my continual refusal to look deeper into the underlying impulses that were shaping the life I was then living.

Once Upon A Time, there was a man who had a pain in his thumb. And over time the pain just kept getting worse and worse. So, finally, in desperation he went to a doctor and explained the problem.

The doctor looked at him a moment, then said, pointing to where the man's left hand rested on his thigh, "You know, if you would stop hitting your thumb with your fist the pain would stop."

The man looked at the doctor in surprise. "I don't know what you are talking about," he said. "I'm not hitting my thumb. What kind of idiot doctor are you anyway?"

I (and I'm sure hundreds of other people) hated hearing that story. But it inevitably comes to mind whenever someone who loves me decides to point out some ineffective thing I am doing and which I am absolutely sure I am not doing anyway and I can prove it.

The diagnosis is terminal. And the first step in coming to terms with a diagnosis is to *accept* it. Then and only then is it possible to figure out what to do *now*.

There are a great many problems that come with the acceptance of a difficult diagnosis, especially one that is terminal. That particular territory is filled with problems, a lot of them. And the major one is that this sort of thing isn't an intellectual exercise. It's a territory filled with feeling, with emotions, with implications that themselves engender more feelings and emotions – and *importantly* these are not quiet, civilized, well-mannered emotions and feelings. This is not a territory of the dissociated intellect. It is a territory filled with grief and loss and confusion and the lonely, lost child inside us calling out for consolation from those who loved them when they were small, and an insistent "no, this can't be true" voice in your head, and rage at the unfairness of it all, and totally unrestrained emotional responses, and fuck you I'm talking about survival now. It's filled with recriminations, and blame, and uncertainty, and terrible, terrible fear. It's the territory of dissolution, the falling apart of everything that was believed to be solid and reliable and endless – including our own bodies and psychological selves and the mostly-virtual but still necessary world that we have carefully constructed around and inside us from the moment we were born.

One of the inescapable truths about all these feelings (which nearly everyone tries to avoid) is that there is no way to deal with what's happening except to go *through* it. It's a *territory* of the self. It is not a place but a moving, ever changing, living scenario in which we are inextricably embedded because of the nature of the diagnosis. It's only possible to come to terms with it by going through what it demands of you as you do the difficult work of accepting and integrating the reality of it. Only afterward can you find whatever life is to be found on the other side. And "going through it" means learning about a lot of things that you never suspected existed and even more things that you never wanted to know about. Not ever.

Still, as with grief and shame and guilt, there are teachings inside a terminal diagnosis, essential truths that can only be found once its inevitability shows up in your life and you finally – after all the screaming and arm waving – come to accept it. And what is really, really true is that a terminal diagnosis will, in the end, find each and every one of us no matter what we do or who we are. For life is, and has always been, the only-and-always-fatal sexually transmitted disease that is or ever will be.

But to make things harder than they ever have been in the history of our species, each and every one of us is getting the same diagnosis at the same time. That diagnosis says that not only are we individually in trouble but also our civilization, our cultures, our ways of being, our science and technology, our ways of thinking, our behaviors for the past two thousand years. They are no longer sustainable. For they were and are based on assumptions about Earth and ourselves that never have been accurate to reality. More seriously, the diagnosis tells us that the way Earth has been (that is, the climate our species has known) for so very long is ending and that it will never, ever come again for any of us. Not ever.

As is true with every terminal diagnosis, the most terrible fact of all has found us: we are facing an absolute and irrevocable ending. The end of everything we have known – individually and as a species. And none of the adaptations or learnings or strategies or skills we have developed over the course of our lives (or during the past two thousand years) can stop it – even if we might have been able to fifty years ago. What is true is that the Earth's ecosystems are collapsing and everything built on top of those ecosystems and the way they have been for so long will collapse as well. (Which they are already beginning to do.) That is just the way it is.

Do you hear that, Mr. Anderson? That is the sound of inevitability.

And with that sound comes a particular feeling, a feeling that only comes with the acceptance of an inevitability that can't be escaped. It's not a pleasant feeling. It is in fact terrifying and it is filled with the sound of weeping.

But we are not, individually or collectively, at an acceptance of the diagnosis, an acceptance of its inevitability. We are instead, as a species and individually, in the midst of struggling with accepting the diagnosis. (And one way to think of all the social disruptions that are going on out there, pretty much everywhere you look now, is that people are responding as groups just the same way individuals and their families do after they have been given a terminal diagnosis.) Despite our resistance, we are being forced as a species to recognize our ecological place, accepting the limits it places upon us, accepting that rationality and science and monotheistic beliefs have left out something essential to our successful habitation of this planet, that maybe all this belief in "progress" thing was misplaced, that maybe some sort of reckoning is now inevitable, that in fact the entire belief system upon which we have based our lives might just be flawed beyond redemption, and that the time for deep interior (and exterior) change is upon us.

These are difficult things to hear; they are difficult things to say. People generally aren't very good at saying hard truths out loud . . . for, well

... lots of reasons. Mostly they are afraid of what will happen after they say them. For one thing they will be breaking incredibly strong cultural injunctions that insist on silence about this sort of thing. That insist we only make happy, positive, hopeful sounds. And every one of us knows that bad things usually happen if those kinds of cultural injunctions are broken, especially when speaking to a lot of people at the same time. So, there's a pressure to keep quiet, and, as well, to keep lying to ourselves, to keep avoiding the truth so we can hold on to the old world as long as possible. Because the truth is that every one of us is scared, no matter who we are or how well or poorly we dress or what schools we did or didn't go to and what university degrees we do or don't have.

Saying these things out loud would mean letting the truth in just a bit deeper than it has been. That will make it more real and the more real it is the worse everything feels. There's a serious and severe depression accompanying that realness, an overriding sense of hopelessness and despair. And then the grief comes, the one that always accompanies endings and the recognition of our mortality and the truth that there are realities here that are far greater than the human and that we can never, as individuals or a species, escape them. There is a feeling of helplessness, too, something that always comes with a terminal diagnosis. And then the anger comes – because we were lied to, because the elites were not good governors when they should have been, because the physicians have been bullshitting us so we would feel better, and because when people are scared they almost always turn it into anger and look for someone to blame. (Nobody likes feeling helpless.) And mobs of angry people are really, really scary.

These are some of the reasons why everyone does their best to avoid saying that the diagnosis is terminal, why all the reports of how bad things are always include some optimistic thing that can be done to stop what is happening, offer up some utopian fantasy or technological fix that will save us. Why every article ends with ... *but wait, there's more, and it's hopeful.*

I am, however, suggesting a different approach. I am suggesting that the emotional territory that emerges as the diagnosis is accepted is important. I am suggesting it be entered, that a descent into the darkness be taken, because on the other side of it lies our salvation. False hopes are not the way out of our personal (or species) dilemma. Accepting the truth of our situation and coming to terms with it is.

Many people before us have gone through this sort of thing. They, too, were given warnings that a storm was approaching. And, as it is with us, the evidence that they were facing serious trouble became more obvious with every year of the lives they were then living. The structures around them were collapsing; it wasn't that hard to see really. But, as always, many people refused to see it and held with all their might to the world as it had been, as if that would make it continue. But there were others who, as difficult as it was, accepted what was coming and began to do something about it. One of those was a young poet in Warsaw in 1944. His name was Czeslaw Milosz and he wrote a poem that he called "A Song on the End of the World" . . .

> *On the day the world ends*
> *A bee circles a clover,*
> *A fisherman mends a glimmering net.*
> *Happy porpoises jump in the sea,*
> *By the rainspout young sparrows are playing*
> *And the snake is gold-skinned as it should always be.*
>
> *On the day the world ends*
> *Women walk through the fields under their umbrellas,*
> *A drunkard grows sleepy at the edge of a lawn,*
> *And a yellow-sailed boat comes nearer the island,*
> *The voice of a violin lasts in the air*
> *And leads into a starry night.*
> *And those who expected lightning and thunder*
> *Are disappointed.*
>
> *And those who expected signs and archangel's trumps*
> *Do not believe it is happening now.*

As long as the sun and the moon are above,
As long as the bumblebee visits a rose,
As long as rosy infants are born
No one believes it is happening now.

Only a white-haired old man, who would be a prophet
Yet is not a prophet, for he's much too busy,
Repeats while he binds his tomatoes:
There will be no other end of the world,
There will be no other end of the world.

Nearly everyone in our time, as was true in europe then, finds it very hard to accept the finality of what is happening. For rosy infants are still being born, fishermen still mend their glimmering nets, and sparrows still play by the rainspouts. Our cars have plenty of gasoline and/or electricity and our supermarkets are filled with food. The television still shows us pictures of happy, comedic families. Advertisements still tell us of the wonderful things we can buy. But what is really true is that our old world has already ended – it's just that its ending is occurring on slower, more geologic time lines than our quickness can easily see.

The reverberations of that ending are moving through every ecosystem on Earth and through every life form there is. Our children, as do the trees and the bacteria and the birds and plants of the fields, *know* it. And they have been telling us about it in the language that each of them knows for years and years and years. Sooner or later will come the lightning and thunder, the signs and the archangels' trumps and perhaps then all will know. But the children already know (they are yelling as loudly as they can). And many other forms of life on this planet know, including those of us who have spent the decades of our lives working for Earth, who have made the difficult journey to becoming Earth looking out of human eyes.

Every human being, in some fashion, in some part of themselves, knows that the time is already here and that there will be no other end of the world.

Nevertheless, knowing this truth in some hidden part of the self is different than accepting its inevitability with the conscious "I" who lives up here in daylight. There is something in the human heart that will not give up optimism, even when the firing squad is lining up behind us.

Things are not yet so dire for most of us as a firing squad lining up behind us – though, importantly, they already are for some. More accurately, for those of us in the west, we are much like a beloved spouse who's been told that our loved one is dying. And here we sit, beside them, unable to accept it. The one we love has had a few pains, yes, but they still function pretty well most of the time, just as they always have. We're still living in the old life, in the old relationship, the one we've known since we were young together and first fell in love. Things are going on seemingly as they always have.

Over time, though, as always happens when life has become terminal, the functioning of the one we love decreases. It's slow at first. But bit by bit by bit, their systems fail. Unexpectedly one day, there's a serious relapse. Then the trip to the hospital comes, the struggle with the relapse, then a recovery but to a lesser level of functioning. Afterward things go on sort of like they did before. We get used to it. Begin to settle into the new normal as if it will continue forever. Then suddenly, the pattern repeats itself. Then . . . it repeats again. And then again. We go to doctors, demand second and third and fourth opinions. Finally, finally, finally . . . it sinks in, there really is a problem that's not going to go away. It's time to get serious. We insist something be done. Certainly our modern, technological medicine has an intervention that will stop this from happening. And yeah, there is (we're told) but there is no guarantee it will work. It's possible that the side effects will be terrible. Nevertheless, we soldier on and insist on the intervention. When it doesn't work, well, let's try the next one and then the next and the next. Anything to keep our beloved, and our old world and way of being, alive.

Sooner or later, though, the truth has to be faced. What we have had for so long we are going to lose. We *are* losing. The one we love is slipping away and there is nothing we can do to stop it. And as the days go by, the tiny filaments "holding the personality to its past shell" are broken – one by one by one. The truth slowly sinks in and finally, we begin to accept it and to grieve.

If we are wise, the two of us grieve together while there is still time.

For the beloved, after all, is the one we have gone to all of our lives. They are the one we needed to hold us when times were hard, when our dreams had failed or been crushed, when we lost something or someone more dear to us than we knew, when the pain became more than we could bear. But now the one we have gone to is the one being lost. When they are gone, who will hold us then? There is no one who truly matters or can give to us in the way that the beloved did. For there is no one that *is* them. And finally . . . it sinks in. When it does we face an emptiness. And loss. And pain. Something essential to who we are is being torn out of life.

I have gone, always, to my beloved, and she has companioned me when my heart was breaking, in times when I was more afraid than I'd ever been, when I did not know how to go on, when the darkness surrounding me had overcome any tiny flickering light I could find within me. And I have also gone, when my heart was breaking, when I had nothing else to hold me or to hold on to, to Earth, to the great forests and mountains, to the wildness of this world, and they have held me in my grief. In that holding they uplifted me and gave me something that can never be found in the human world, something that all of us, no matter how dissociated we might be, living in our cities and apartments or suburban homes, need more than we know. In that holding I found the strength to go on, the wisdom I needed to continue to live, to work to bring the gifts I was given into the world in some tangible form. But sometimes now, when I go to the forests I have known all my life, there is nothing there but broken stumps and endless plains filled with emptiness. When I go to the mountains, too often I find there only open pit mines and men and women smelling of evil, "inquiring into the purchase of our homes." And maybe it is the same with Earth as it is with my beloved: if I am wise, I will reach out so that we can grieve together while there is still time.

We are intertwined with the world; we form more bonds than we know. *Grief will be our companion on this journey – it is not something we can deal with and move on.*

INTERLUDE FOUR

Fragments from a Stained Glass Window

For among these winters there is one so endlessly winter that only by wintering through it will your heart survive.

RAINER MARIA RILKE

We did not come to remain whole. We came to lose our leaves like trees. Trees that start again.

ROBERT BLY

I know of a country that spiritual flatness does not control, nor constant depression, and those alive are not afraid to die.

KABIR

"What happened then?" I asked.

It was nearly midday; the five of us had just gotten out of bed. We were sprawled on the couch and chairs in the living room, cradling cups of coffee, newly lit cigarettes in our hands. The place was awash in beer bottles, most of them with a cigarette butt or two inside, food-encrusted plates on every flat surface, a few half-smoked joints here and there – the usual mess that's left after a group of twenty-somethings had spent the night partying. We were in the high mountains of colorado in 1972, the world was new, and we had our whole lives before us.

Billy was blushing a bit, laughing in spite of himself, as he told us why the girl he'd spent the night with had already gone home.

"Well," he said, "I only had the one sleeping bag and it's a single. Somebody had already passed out on my bed, so we were both crammed into it, her on the inside. It was cold, so I'd zipped the thing up. We were almost asleep when I farted."

We groaned... then began to laugh as the scene played itself out in our minds. Billy had the worst gas of anybody we'd ever met. It was always silent and very, very deadly. We'd suffered it ourselves often enough.

Billy started to laugh, too, as always irrepressibly childlike. "Well, a minute or so later I heard her go, '*Oh, God! . . . . . . Oh, God! . . . . . . Oh, God!*'"

"What did you do then?" I managed to ask.

"I pretended I was asleep." And then he laughed even harder.

By then the image was so clear in our minds, our own experiences fueling our imagining, that the rest of us were laughing uncontrollably as well. That poor woman, trapped in that bag, with Billy on the outside and no way to get out.

"Billy, you're incorrigible."

Despite his deadly emissions, Billy was an endless source of amusement to all of us. It was impossible not to love the guy. He'd retained the child of him more than anybody I'd ever known. I just took it for granted though – as I did so many things then, before I was able to understand how incredibly special those people and the adventures I had with them were. Like the time all of us, excepting David, decided to make beer. (David was, as he put it, into drinking, not making.)

Hal was predictably excellent, brewing some very fine ales over the years. John had a go once or twice with decent success and I, for some reason, really got into it. So much so that years later I ended up writing that book about it. And then there was Billy.

Hal had explained to him, in exacting detail, just how to go about it and Billy had (we all thought) paid close attention to him when he'd done so. Eventually he borrowed Hal's equipment and brewed up his own batch. He bottled it when the primary fermentation was done, priming the bottles before capping them of course, and waited the necessary few weeks for the carbonation to build up. Then, one afternoon, with a great flourish, he brought the beer out for us to drink.

We were sitting around the living room again, in that rambling old house where they lived then – the old stage stop on the Golden Gate Canyon to Central City line, a resting place for horses and passengers before the long climb up to the mines. Billy opened the first bottle for himself, then passed around the opener and we each popped the lid off our own. Then carefully, to lessen the head, we poured the beer into our glasses. The most beautiful amber beer I'd ever seen caramelled itself into our glasses . . . followed by cigarette butt after cigarette butt.

"Billy!" we exclaimed.

"What?" he said, truly confused. "You're supposed to wash them first?"

We couldn't help it. We fell into laughter again . . . at the look of honest confusion on his face, at his innocence, at the child of him that we so loved. And it made perfect sense, too, when he took that first sip, then said, "Mmmm, with just a hint of smoky flavor on the back of the tongue," and drank it down. Then he laughed and laughed and laughed.

It was so easy to fall into friendship then – just as we had when we were children. We eased into it as if it were the most natural thing in the world. And we found a whole world inside the friends we made: John, the ultimate norwegian bachelor farmer (he would never marry). Hal, the charismatic intellectual, whom we believed destined for great things (a master stonemason and martial artist as it turned out). And David, who had the saddest eyes I'd ever seen, yet leavened with a melancholy *gravitas* that affected me deeply. And me, caught for so many years in the woodenness of my life and only friends like this to see me through.

Hal was a temporary tenant, waiting for his beloved to come to colorado to be with him. She needed another year in ohio but he was patient . . . and very much in love. When she finally arrived they found a place not too far from my little cabin in the canyon and I'd visit them from time to time.

Once, as we sat on his porch drinking beer, Billy told me that she'd been in an intense, passionate love affair before she'd met Hal. When the guy dumped her she'd shot herself in the stomach. She'd been in the hospital a long time before she recovered and I always fantasized that was when Hal fell in love with her. After she healed she finished her degree, did her final year of training, and came to him. She was a social worker now, counseling women in difficult relationships for the most part.

I used to wonder what her stomach looked like. And what it was like to recover from something like that. She was elegant and pretty and a therapist and saw into me and in ways that scared me and I was shy and tongue-tied and couldn't talk to her and I don't think we ever had a real conversation in all the years I knew her. But I've never forgotten her ... or the story that Billy told me.

Maggie was late. She'd just gotten off work. And while the restaurant was only a fifteen-minute drive from her apartment, the managers had kept her and a few others in the office after their shift ended. They'd just been to a week long workshop for business owners and one of the maxims they'd learned was "Being on time means not being late," which they kept repeating until Maggie just wanted to scream: "It was only five fucking minutes!" And so, here she was, running late, all because of that fucking workshop. It was going to be a challenge to be on time to her therapy group, a good hour south of town.

She'd been going for a few years now and had been, slowly and with much resistance, descending into what her therapist kept calling the "primal survival drives and fears" that lay at the root of her difficulties. They were layered into pre-verbal behavior patterns that she'd learned from her parents. And her parents ... well, they'd been pretty fucked up actually – though it took her a long time to see it. All those times they'd locked her in the closet when she spilled food on her dress, well it just seemed normal ... until she talked about it in group one night and her therapist pointed out that it was actually child abuse. Still, the deeper she went inside herself, the more scared she got, as if something down there just didn't want to be seen.

She was thinking about all this as she rushed home. (Therapy night always brought up a lot of thoughts she didn't want to have.) She lived on the second floor of a three-story victorian house that had been converted decades ago into apartments. Maggie pulled into the parking lot in a spray of gravel, jumped out, and ran around the side of the house to her entrance. She unlocked the door at the bottom and ran up the enclosed stairway to the landing, where she took off her shoes. Then she unlocked her apartment door, stepped down so the outward-opening door would clear her feet, and went in.

She threw her keys and purse on the kitchen counter, went into the bathroom, and turned on the bath water. In the bedroom she shrugged off her clothes and pulled out what she was going to wear. The blouse she really wanted was in the dirty-laundry bag in the car so she found a different one and laid it on the bed. Her hanging-around shoes were still on the outside landing where she'd left them the night before. Naked, she ran to the door, opened it, stepped out and down, reached over to pick up the shoes . . . and heard the door swing shut and lock behind her.

There really isn't anything quite like the feeling that comes when something like that happens. Being naked on a stairway just as evening is falling, keys on the counter inside, a locked door, and bath water running full out. Her mind just went blank. She froze, feeling as if she were caught in amber, something future generations would find there on the landing, a human insect trapped in time. And somewhere in the back of her mind she heard a low, satisfied laughing.

She tried to move, to begin to think again, but it felt as if a tremendous force were impeding her. A great many old thoughts returned: how useless she was, how unlovable, how inadequate, so clumsy. And still she stood on the step, immobile, frozen. Then somewhere inside herself she began to get mad. As it welled up inside she heard her therapist's voice: "Anger is just energy to solve a problem, nothing more." So instead of repressing it as she had so many times before she just let it get stronger and stronger.

As the anger moved upward through her body the stasis in which she was held began to break. Finally, she shook herself, ran down the stairs, cautiously opened the lower door, and peeked out. Making sure the lock was *not* engaged, she quickly snuck through the yard to her car. Grabbing the bag of laundry, she pulled out some clothes and quickly dressed. Then, opening the tool box on the floor, she grabbed the hammer and screwdriver. She ran back around the house, up the stairs, and got to the landing just as the phone began to ring.

She used the hammer and screwdriver to pop the hinge pins up and out of the hinges. Then she inserted the screwdriver into the crack between the hinge side of the door and its frame and wiggled the door out. Dropping the screwdriver, she pulled the door out of the frame, turned it sideways, and carried in into the house, where she leaned it against the kitchen counter. She ran into the bathroom and turned off the water just as it reached the top edge. Then she hurried back into the kitchen and answered the phone. "No, I can't talk right now. I'll call you tomorrow."

She took her bath, dried herself, then ran into the bedroom and put on her clothes. She picked up her keys and put them in her pocket. She got her shoes from the landing, went back inside, and put them on. Then she picked up the door and rehung it in the frame. She stepped back inside, grabbed her purse from the kitchen counter, stepped out again, locked the door, ran down the stairs, out to her car, and set off for group. Just in time, she pulled up in front of her therapist's house, got out, and, taking a deep breath, walked slowly up the sidewalk and opened the door. As she did, she felt a part of herself scurry off into the darkness of its hidden room and lock the door.

She paused a moment on the threshold, looked inside, and thought, "So that's what was going on." Then she straightened and walked into the room and said, "Hi, everybody. Did you know that being on time means not being late?" and began to laugh.

Once Upon A Time, in a world very much like this one, there were two young men who were driving an old, battered car on a narrow road late at night. They'd heard tales of an indian medicine man who was teaching the way of Earth to any of good heart who wished to learn. They were pretty sure they had good hearts (though sometimes they wondered if that was true – they weren't really sure what a good heart was) and something about it seemed to call to them so they gassed up the car and took what little money they had and set off on a journey. To find themselves I guess or maybe to find a way to be on this Earth that felt a little better or perhaps to learn something about the soul if they were lucky – though this was just a feeling they had, something they never let themselves say out loud, even in the privacy of that huge dark place they carried inside.

They had driven through the heat of the day, then toward the lowering sun as night began to spread its blanket over Earth. The moon wouldn't be showing her face this night and so, before long, the darkness around them was very deep.

The road they drove was old and worn and not much traveled; they had taken a path that few now thought worthwhile. The headlights were dim and yellow because the car's generator had seen better days and of course to make things more difficult the two young men were seriously stoned and had been driving for a very long time without stopping.

They were talking as young men often do of some things that really were important and many others that time would reveal were not when suddenly a coyote ran right in front of the car. There was a sudden whump, then a silence. Trembling, the driver brought the car to a halt. The two guys looked at each other, uncertain and shaken. Then they got out of the car and slowly walked back down the road. The coyote's body lay on the edge of the pitted asphalt. Cautiously, afraid, they approached it. After a moment, the driver carefully reached out his foot and gave the body a quick push. Nothing. Again. Nothing. The coyote was dead.

They stood awhile looking at the body (they were *really* stoned and it took a while for thoughts to form) and eventually they began to talk. They were sure it was a sign. They knew they were supposed to do something but they weren't sure what. (They were very new to signs.) Finally they decided that they were supposed to take the coyote to the medicine man they were on their way to see. So they loaded the body into the backseat of the car and, still a bit shaken, they set out once more driving that long, lonely road into the darkness.

Deep philosophical conversation followed. They went over and over just what this sign could mean. They were sure that Coyote was in some way blessing their decision to study with the medicine man. They knew that one of the man's totem animals was Coyote and they spent a good hour discussing the ramifications, going over the spiritual meaning of Coyote and the soul reasons for His emergence at just the exact moment when they and only they were driving on that road. They could feel

Creator moving purposely in their lives, blessing them and this path they had decided to take. They were pretty sure of this especially since they had intentionally decided not to take the interstate, which all normal people do, and instead followed the old highway, the road no one traveled anymore.

The more they talked, the more special they felt, for surely the Great Beings of the Universe had stepped into their lives. And the more they talked the more grandiose they became. They were pretty sure they were going to be the most special students the medicine man had ever had. Images emerged in their minds of how amazed he would be when they drove up and gave him the coyote, of all that he would teach them then, all the ways they were going to change the world, all the ceremonies they would learn to do, and the thousands of people who would come and work with them once they had.

Then, during a pause in the conversation, they became aware of a slow, steady growling coming from the backseat. The passenger slowly turned his head as the driver looked in the rearview mirror. Standing unsteadily on the backseat, weaving with the movements of the car, was the coyote, yellow eyes glaring, teeth bared. He seemed pretty mad actually. There didn't seem to be any particular spiritual meaning for him in all of this – just an unpleasant encounter with a couple of stoned young men who had not been paying attention as they drove a deserted highway late at night.

They screamed. The driver took his foot off the gas and the car began to slow. They opened their doors and threw themselves onto the road. The car, doors still ajar, slowed more, wobbled off the road onto the grassy side, and shuddered to a halt. The coyote hopped over the seat, then out the front passenger door, and limped off into the night.

Eventually the young men got back on their feet. They were shaking from adrenaline shock, their arms and hands bleeding from road rash. Stumbling, they climbed back into the car and once they were a bit calmer, set out once again, much more slowly than before.

It takes a long time to learn the meaning of signs. Grandiosity is always a problem.

I go to Upaya Zen Center in Santa Fe, new mexico sometimes though not very often. The temple that Joan Halifax built there is one of the most beautiful I have seen. (And I have seen many that have been crafted by those who know the language of wood.) She'd hired artisans that knew how to build in the traditional way, using the same ancient, time-honed tools that had always been used in the building of zen temples.

The room had come alive as they worked, simply because of the love they put into it and all the talking they did with the wood as they called it to awaken from its sleep. And then, too, it's grown ever more alive from all the sitting meditation that has taken place in the decades since it was built. All that love and all that focus have penetrated deep into the wood of that living, breathing room like some sort of ancient preserving oil. And it's brought out a patina that I think can occur no other way. With each passing year the room seems to come *more* alive and aware. Its powerful eyes looking deep into everyone who steps inside it, into the interior of a living intelligence that rationalists and fundamentalist skeptics insist does not and cannot exist. Into what truly is a temple.

I was there this night to attend a sitting practice and hear a lecture from a visiting sensei. As the evening began, there was some initial talking and settling to be done, the usual announcements (always boring and always so mental). Then the great bell made its sound and into that reverberating, ebbing silence, we began shifting into a different state of mind.

Sitting meditation, in a room with a hundred practicing buddhists, is a tremendously powerful experience. It's always a bit scattered at first, people's thoughts scurrying this way and that as the thinking self tries to find shelter from what's about to happen. Slowly, inevitably, breathing deepens. And as it does, everyone drops deeper into themselves. Soon, the quiet comes, a quiet that slowly but surely fills the room . . . and the spaces inside every person there with a resonating, living meaning that is nearly impossible to capture in words. Then, at the moment it reaches fullness, in just a quick instant of time, the hearts of a hundred people suddenly entrain with each other. Synchronized, we all became, for a time, one living *experiencing* organism, held in the living field and intelligence of that sacred room.

The state that is found, and in which we were sitting, is not something that can easily be captured in the language of the mind. It is a nonrational state of being where we become, in that deep silence and quiet breathing, experiencers of the experience of being. And as always when I am in that state, I feel the touch of the living Earth upon me. Its powerful, ancient movements filled with meanings that humans can only barely grasp. Those meanings have been building themselves for billions of years. They are generated from the contemplation of Earth, the organism from which all of us have come and that feels, when touched by human awareness, as a great, ancient redwood does but possessed of a power a million times stronger. It is the elder of elders and it has been growing and contemplating for a very long time.

The buddhist path is not mine but I visit there every now and again. I come instead from the wildness of the world and remain, as I always will, undomesticated, uncivilized. Not of the cities. I am *vegetalista*, ecstatic animist, not buddhist. Still, I like the zen world very much, just as I do so many of the people I have met there. Sometimes, it's fun to hitchhike along with them as they travel, as they move into nonrational states of being and mind. For the *vegetalista's* path is often a lonely one. We travel between the village and the deep wilderness of the world (living fully in neither) and bring back what we find for the human community. We are People of the Plant and it is at the foot of the Green that we sit. It is the Green who has taught us the Way. But most always, we travel alone.

I have spent a half century in the practice that has been taught me, forms of meditation and contemplation that belong solely to the *vegetalista's* path. (Though, of course, there are overlaps with the path that others travel for all of us are headed toward the same destination.) There is no way to survive this path or the journey it demands without a powerful, deeply ingrained practice that is followed each and every day of life. (And again, a practice can be *anything* at all, really, from stamp collecting to the shaping of wood to sitting meditation. It is not *what* you do but *how* you do it that matters.) Like all *vegetalistas*, I travel into multiple states of mind and being as a way of life. But sometimes it's nice to have company. And besides, the scenery along the buddhist path is very lovely indeed.

As the field synchronized, and we entrained more deeply, I could feel how powerfully that potent, entrained state of being flowed into the depths of us. And I understood as I looked around the room just how it is that sooner or later zen practice breaks open habituated patterns of thinking and behavior in the people who follow this way, how and why it breaks open the entire assumptive world that people have inside them. And why sometimes it is the people themselves who break as it does.

Although my path is a different one, it's fundamentally similar in some important ways. For both my path and theirs travels into the nonrational world. We walk a balance beam as we do, the way is very narrow, and one can fall this way or that if care is not taken. And, of course, all of us do fall, many times, until we learn that most necessary of skills. Balance. This is one of the great truths that few people easily learn; I was not one who did. For the only way to learn balance is to lose it over and over again. And when we fall? Sometimes the pain is terrible indeed.

But that state of entrainment? It emerges of its own accord but it can only be maintained because of the balance that has been learned over long years of difficult, continual practice. And as the self moves more deeply into it, ever more fully into the nonrational world, psychological clothing falls away, cultural clothing, species clothing – they all fall away. The soul opens itself outward then, taking on its original shape and for a time it takes flight in the world. It's truly a wonderful thing, to look outward from the soul's eyes and see another soul looking back. What a wonder it is when two souls touch in this way.

It is worth the price that the skill of balance demands.

And so, for a while, one hundred of us sat in that marvelous state of being. Something in us, in our souls, was renewed as we did. In some strange fashion we knew more clearly who we were again, what our purpose was, felt renewed in our allegiance to this path we have chosen – or which has chosen us. Then, as it always does, as it always must, the field in which we were embedded shifted, began to break apart. Our minds slowly started their chatter again, and once more we began to move into our social, daily ways of being. Nevertheless, echoes of that nonrational state still resonated in the room . . . and inside us. And so, for a while, we would remain partly in one world, partly in another.

Into that moment, there in front of us, the speaker for the evening seated himself. His robes were perfectly composed, his face serene. Then he began to speak. And that state of mind, still so strongly present in us, seeped away like water into porous ground, going back to wherever it lives when human beings tilt the balance toward mind alone.

It was only when he'd finished his prepared talk (and yes, it was terribly boring) and opened up to questions from those of us in the room that he, and we, began to come alive again. There were several engaging questions to which he responded with wit and elegance. He parsed some really tricky metaphysical points quite well, revealing as he did just how deep his thinking had been over the years. Still, it is the last questioner who has stayed with me – as I suspect he always will.

He was a young man, in his mid-twenties, a social justice and environmental activist. He was in terrible pain, of soul, of mind, of heart. And his question went something like this:

"I work so hard for the Earth but there's so much damage – in the Earth, in people. I see longtime environmental activists working all day to protect the rain forest. Then they go to McDonalds for hamburgers (and here his voice broke, filled with pain, and anger, and the weeping of centuries). I see all the plastic pollution and then watch activists buy and drink bottled water. (Again his voice breaks.) There are so many problems, we're just surrounded by them, caught up in them. They are everywhere. Every place I look there is damage and pain and my question is how do I deal with all this, what do I do in the midst of this pain? There is just so much to do and I feel so overwhelmed, more and more overwhelmed every day. I don't know what to do." And he began to cry then with those terrible, shuddering sounds, the ones that come with the breaking of men who have been trained not to weep.

The speaker sat a moment, just holding the question suspended inside him, allowing it to hang there in the room until it had fully penetrated the depths of us, waiting until fullness was reached. I saw a look on his face then, one I recognized. It's one I've seen in my mirror many times over the years. I looked around the room and saw it, too, on Joan's face, and on many of the older activists' faces as well. For each and every one of us have faced this same problem and this same question. It's a terrible problem to have inside the self and it's a terribly difficult question to find an answer to.

The sensei finally responded and it went something like this . . .

"We do what we can and that is all we can do. We walk in the midst of pain, of the suffering of the world, every minute of our lives but we cannot rid this world of that pain even though we each try in our own small way to help end suffering. It is and was and shall be. We can only alleviate it. If we try, by ourselves, to heal all of the troubles in the world it will overwhelm and destroy us. We will break under the load.

"Over time, I have learned that it is necessary to find the area or areas I am most fitted to address, that by my nature I can help with the most. That is where I put my time and love and work. Afterward, I still walk in the midst of the pain, of the troubles of the world. But I know as I do that the pains and wounds I cannot respond to must be addressed by other people, that, in fact, there *are* other people addressing them. *There is an important faith in this.* All of us are struggling, none of us have gone far. But it is crucial to remember that there are others who feel as you do and who are doing the work. None of us are wise enough, or strong enough, to solve all the pain, all of the problems, in the world.

"The hard thing is to let go of what you cannot do and trust that others will respond to what you cannot or are unable to do. And it's difficult, I know, to learn to walk in the midst of that pain without losing yourself. We, each of us, have to learn how to endure the pain so we can stand in the midst of it and remain ourselves. Our personal task is then to work with the suffering that calls to us the most strongly. That is what you must focus on, that is where your soul work is to be found. The rest is up to other people and other forces in the world."

I could tell that the young man did not really understand what the speaker had said, but I saw him wrap it up carefully in his heart cloth and put it away so that he could find it in the years to come. I knew that he would take it out, unwrap and contemplate it, many times in his life – as I have done and as every person in that room has done when touched by truths they did not yet understand. And I knew that one day, he would indeed understand it – that is, if he survived his descent into this particular darkness. For there is no guarantee that when we – any of us – make this particular descent we will survive. We must learn, as Robert Bly once said, to grow our wings on the way down.

Those of us who have been called to respond to the suffering of the world learn over time that we are limited in what we can do. The world is filled with suffering and no single set of shoulders can carry that load. So, over time we find where our true chance of greatness lies and there we take our stand. It is then that we begin our life of service.

William Stafford often wrote of what is involved in this, as he did in his poem *A Ritual To Read to Each Other* . . .

>*If you don't know the kind of person I am*
>*and I don't know the kind of person you are*
>*a pattern that others made may prevail in the world*
>*and following the wrong god home we may miss our star.*
>
>*For there is many a small betrayal in the mind,*
>*a shrug that lets the fragile sequence break*
>*sending with shouts the horrible errors of childhood*
>*storming out to play through the broken dyke.*
>
>*And as elephants parade holding each elephants tail,*
>*but if one wanders the circus won't find the park,*
>*I call it cruel and maybe the root of all cruelty*
>*to know what occurs but not recognize the fact.*
>
>*And so I appeal to a voice, to something shadowy,*
>*a remote important region in all who talk:*
>*though we could fool each other, we should consider –*
>*lest the parade of our mutual life get lost in the dark.*
>
>*For it is important that awake people be awake,*
>*or a breaking line may discourage you back to sleep;*
>*the signals we give – yes or no, or maybe –*
>*should be clear: the darkness around us is deep.*

I read this poem, sometimes, late in the night, when I wake and my soul is at its lowest ebb. And then I remember the people I have met and the roads I have traveled and the poets and teachers who have helped me on the way. Sometimes I walk out into the night and look up at the darkness of the world, at the stars in their billions, and feel what it is like to be an expression of this Earth at this time in the history of the world. Somehow, all these things help and it's then I find my way again. I don't feel so alone. For what is true is that others have traveled this way before me. They are not so far away as they seem to the daylight mind, or to the soul when it awakens in the depths of night. Those others are often very close indeed. And they will reach out to us, helping as they can, if we will only let them do so, if only we will ask.

CHAPTER FIVE

The Journey Through Grief and Loss

*Depression is a reaction to grief over which nothing can be done
... and it comes with a sense of helplessness — that whatever you
do, it's impossible to bring that good relationship back.*

CIARA O'ROURKE

*Coming to trust the darkness takes time and often involves many
visits to this land. Our arrival here is rarely a chosen thing. We are
thrown into the darkness or are carried there on the back of a blue
mood. What we make of this visit is up to us. Recalling that the darkness is also a dwelling place of the sacred allows us to find value in
the descent. In this place of lightlessness, we develop a second sight.*

FRANCIS WELLER

*Those who recover from bereavement do not return to being the
same people that they had been.... Nor do they forget the past and
start a new life. Rather, they recognize that change has taken place,
accept it, and examine how their basic assumptions about themselves and their world must be changed and go on from there. Each
of these steps require courage, effort, and time. Three distinct steps
in recovery are: first, that the loss be accepted intellectually; second,
that the loss be accepted emotionally; and third, that the individual's model of self and outer world change to match the new reality.*

COLIN PARKES

The dangers of life are many, and safety is one of them.

GEOTHE

Hiding in the shadows underneath the grief of ecological loss is a truth that this book has been written to reveal – to both the feeling heart and thinking mind – so that you, who are reading this, can begin to get a *feel* for it. So that you can perceive it with your conscious mind in the full light of day. So that, with every part of you, you can more easily contemplate and begin to come to terms with it. And that truth is that an end is coming, that in substantial ways it is already here, and that there is nothing that can be done to stop it. For it is only when the *ending* that our feelings of ecological loss and grief are telling us about is accepted as a real and true fact that it becomes possible to find the way of being that awaits us on the other side of our grieving.

I have heard rationalists say many times that of all the species on the planet, only we (human beings) know that we will die. I am not sure *how* they know this for I am pretty sure they don't really know what goes on in the interior lives of all the kindred species with which we share this planet. But for sure, they don't really understand what goes on in the interior lives of people either. Because, despite the fact that everyone has heard the *rumor* of our individual, sooner-or-later-can't-be-avoided death, not one person actually *believes* it will happen to them. Everyone is pretty sure that an exception will be made in their case. That is why, at the end, it takes a lot of illness and debility to convince people that no exception is going to be made. That it is, in fact, their turn now.

I think that this is especially true in western cultures because death has been so carefully hidden away from us and so many promises have been made that we can escape the ecological realities and limitations that govern us. And so, when death comes calling, the denial is so strong that it's very hard to come to terms with what is happening. Acceptance takes a long time. ("I am not meat. I am not prey. I am not meant to be eaten. I deserve to be safe.")

I suspect that at this moment, as I say the world *death* out loud, you are poignantly aware that it is a word very few people in america feel comfortable saying. (I am breaking a social rule to do so; I suspect that you feel some of the emotional impact of that breaking as I do.) This is why americans rarely die but instead "pass away" or " take the long sleep" or "transition" or "find a better world" or "join with the angels" or "is with god now or "once again become stardust" and so on and on and on. People do a lot of things now at the end of their lives but they don't die, especially in america. And for sure they don't die of old age. They die of smoking or eating the wrong food or drinking alcohol or not exercising enough or some other thing and if they had avoided [*whatever it was*] they could have lived forever and ever, amen.

Even when people make out their wills, assuming they can get over the emotional hurdle of doing so, they only do it because "something might happen." And we all know what that means, don't we? (The word *death* whispers itself inside when we hear that phrase, but softly, softly.) Yet it's socially prohibited to say it more clearly, loudly. Why? Because we live in a country that has made denial its national obsession. And that denial, as Robert Bly puts it so well, is a mark of "the naive person's inability to face the harsh facts of life" and, as he says . . .

> *The health of any nation's soul depends on the capacity of adults to face the harsh facts of the time. But the covering up of painful emotions inside us and the blocking out of fearful images coming from outside have become in our country the national and private style. . . . denial begins with the refusal to admit that we all die. We don't want anyone to say that. Early on in the cradle, swans talk to us about immortality. Death is intolerable. To eat, shit, and rot is unthinkable for those of us brought up with our own bedrooms. We want special treatment, eternal life on other planets, toilets that will take away our shit and its smell. We love the immortality of metal, chromium implants, the fact that there are no bodily fluids in the machine, the precise memory the computer has, the fact that mathematics never gets colon cancer, and we are deeply satisfied that Disneyland can give us Germany, Spain, and Morocco without their messy, murderous, shit-filled histories.*

This deeply ingrained denial is another of the reasons grief does not make its appearance in the books, articles, and journal studies about ecological loss and grief that I have read and sometimes quoted in this book. Death denied will not permit grief to be felt in the fabric of their words. Those who are writing about climategrief/ecologicalgrief/solastalgia have not accepted the implications inherent in the pervasive feelings of ecological loss among the world's peoples. Yes, the forests are being cut down, and yes, a lot of other bad stuff is happening, and yes, we feel the impact of it but (for those writers) all we have to do to deal with it is to talk *about* the grief, or maybe go to a therapist to get some help, and for sure we should not be alone in the grief, maybe we should take some antidepressants to get us through or be in a support group, or perhaps become an activist of some sort. And for sure we *don't* have to talk about the death and endings that those feelings of loss are telling us about because, of course, the human species will do something in time to ward it off. It is not going to become real. Death is not going to happen. We can all go on just as we have been going on, forever and ever. And life will get back to normal. Because for sure we need normal now more than anything and that is what you have to put your faith in.

But what is really true is that when death comes calling, life does not get back to normal. Not ever. And I am pretty sure that a lot of people who are feeling Earth Grief know that death is coming, that in many respects it is already here, even if they don't know what to do about it.

It is unlikely that academics and the media will be able to help those of us who feel Earth Grief so strongly in our lives. But there are people who can. They are the ones who have spent time with the people and families whom death has come to visit. They know what happens to a human being when a terminal diagnosis is given. They know what happens to a person when the one they love the most in all the world dies. They know what it does to the structures of the self when someone loses their beloved, the one woven into the deepest part of their heart, identity, and soul. They know the territory a person enters when death comes calling them, when the end of everything their life has been structured around occurs. They are the people who work with the dying, who travel into the territory of death and irrevocable endings, who work with the families of those who have died, who work with people who have had everything taken from them and must still find a way to go on.

These are the ones who understand the landscape that I and you and so many others are entering now. And it is to them and what they have found there that I have turned for knowledge of the way. For no matter what kind of death has found us – that of career, or financial stability, or home, or country, or the Earth's climate, or the landscape in which we have lived our lives, or our child, or our beloved, or our health to the point where we ourselves are dying – the impact on the structure of personal reality is the same. It only differs in degree and sometimes not even then. But still, here we are, the world in shambles, ecosystems in ruins, a way of life we have taken for granted coming to an end.

In that ending there will not only be the loss of the Earth climate or landscapes we have known all our lives, there will be actual deaths as well. There will be the deaths of kindred species that have companioned us for millennia. There will be the deaths of people, too – as the Covid-19 pandemic has made plain – people whom each of us love and care for. And there will be the death of a way of life that we have grown up with and believed to be immutable, unending, permanent.

Grief will be our companion on the journey *because* of what we have lost and *because* the losing will not end in our lifetime – nor will it end in the lifetime of our children or our grandchildren, or their children either. We live inside both loss *and* a terminal diagnosis that tells us that other losses are coming. We cannot simply grieve what has been lost and go on, for there is no normal world to go back to and in which we can go on. The losses and the deaths are not going to stop. Things are going to get worse.

What you who feel the grief of Earth are being asked to do – by Earth, by the Green, by your kindred species, by the ecosystems that are falling apart around you – is to give up denial as a way of living, to turn your face and heart to the "harsh facts of life." That is, to look directly at the death and dying that accompanies what is happening and to what happens inside you as you do. For it is only by facing these harsh facts of life and coming to terms with them that the other side of grief can be found. Importantly, the individual journey we each must make, it turns out, is identical with the one that our species must take if it is to successfully adapt to the world in which it will now live.

And yes, I do know how hard it is to face these things. The stories I have told in this book mark the stages of my own particular journey into "facing the harsh facts of life." It has not been an easy journey for me. I have wept deeply on the way and also felt great pain. I have often felt lost. There are times when I have not known how to go on, not known how I would make it through another day.

The journey has demanded deep exploration in the hidden reaches of my heart, the facing of things I had no desire to face, the discovery of truths that I had never suspected existed, the confrontation of intellectual and rationalist beliefs that were and are inaccurate to the real world, the giving up of other, deeply cherished beliefs, continual, rigorous self-examination, and most of all the willingness to feel the grief that all these things demanded of me.

But I have told you about these things in the way that I have because I think they might have relevance to you and your life. That you might be struggling in the ways that I have, that your ways of denial might be similar to mine, that there might be a commonality between us simply because we are both human, both facing the end of things, the loss of lives and places we love. I have thought, though I can never know for sure, that what has occurred in the deepest regions of my own heart as I have struggled to find my way to the other side of Earth grief might be similar enough to what you are going through that what I say here will help you during your own journey. I hope so. For that is the reason I have spent so long writing this book for you.

There are many aspects to this journey, so many pieces, and so many learnings. Many of them have been found in unsuspected places. Like the time, thirty years ago, when I was sitting around a table, eating lunch with some friends. The two guys sitting with me were also remodeling carpenters (as I have been off and on all my life, simply because I love the language of wood and the art of fixing broken homes). The three of us thought we were pretty tough and it is true that the work we did often demanded both strength and courage. Sometimes we had to lie on our backs and scooch deep into tiny, claustrophobic spaces – the ones that lie far underneath massive buildings, with only an inch or so of space above our faces. And of course, we worked with dangerous machinery every day, machinery from which we could not remove our attention without terrible consequences. And sometimes, we had to stand on wobbly scaffolding against high-storied buildings repairing fascia as strong winds blew. There were the decades of lifting heavy things, the pouring of cement, the kneeling and laying of tile, and the damage that kind of work does to the body over the decades. And every so often, as has always been true for those of us who do this work, injuries occurred. Sometimes terrible ones. Somehow, on that day, we got on that topic and began sharing some of what we had seen happen to others.

The three women sitting with us didn't say a thing. But the more we talked the more they took on a state of being I'd never seen in them before. There was a calmness that came, a strength, a settledness of being if that makes any sense. We had overlooked the fact they were nurses, one in ICU, another in ER, and my friend Sue in OR. When we'd finished, with a clever look between them, they began sharing their own stories. Within a few minutes our faces were white and we'd quit breathing; we looked like little kids who'd suddenly found themselves among grownups, completely out of our depth. When they'd finished, the women shared another look between themselves, and then . . . they smiled, a secret, knowing smile, one privy to secrets that had escaped us in our sheltered lives. They were the ones, not us, who had turned their faces and hearts to the "harsh facts of life." They were the ones who had learned to look directly at death and the dying that always precedes it.

Years later Sue told me another story about her work. It's one that will always live inside me; it's not something that once heard can be forgotten. She was in OR, a boy of ten or twelve on the table. During the operation he'd suddenly gone into cardiac arrest. The team worked hard to save him but they could not. Eventually, the doctor called it (as they say – they rarely, almost never, use the word *death* or *died*. Then they tell the parents things like: "I'm sorry" or "We couldn't save him" or "We did everything we could.")

The team finished what they were doing, the monitors were turned off, and they slowly filed out of the room. But Sue stayed there, silent and still by the side of the boy, her hands resting lightly on his arm. Before long a nursing supervisor came in and told Sue to leave and get on with her work. Without turning Sue said, "I just need a minute."

The supervisor was offended by this of course and insisted quite forcefully that Sue get on with things. Finally Sue turned to her, looked her in the eye and said, "A little while ago I was holding this boy's beating heart in my hands. I felt it stop. I felt him leave. I need to stand here a minute and talk to him and tell him goodbye and that we did all we could. He deserves that. I deserve that. I need to do this before I can go on."

Well, the supervisor didn't take well to that. Not at all. Later she wrote Sue up for insubordination. Nevertheless, that day, Sue took the time to say everything *she* needed to say to the boy who lay there still, silent, and unbreathing on the table. More, she said everything that *the boy* deserved to have said to him as well.

I have always thought her very brave.

I think about that story from time to time for there are so many teachings in it. We have abandoned so much of our humanness simply because we cannot allow the texture of death and loss into our lives. There are so many of us now who no longer know how to behave or be when it comes. Even doctors, even nurses.

But I believe, I have always believed, that part of every physician's duty is to journey with the dying as far as they possibly can, to journey into the grief and fear that is part of that journey and to do so as well with the families of the dead after those they love are gone. And I believe that this is the duty of every kind of physician there is, even nurses, even herbalists, even ecologists, even social justice and environmental activists, even those of us who love our children.

And I believe, too, that it is part of our task now, given the times we find ourselves in, to sit at the feet of death and allow it to teach us. For we live in a time of endings, when many things we love will die – ways of life, kindred species, Earth climate and landscapes. We should not travel into that future unprepared. And so, I believe that it is time for us to sit at the feet of death and allow it to teach us the things we need to learn to make it through the times we now face. For death is and always has been one of the great powers of the world. It has been around a very long time, far longer than our species. It is one of the greatest teachers any of us will ever have.

So, I will talk now about the territory in which we who are dying (as I myself am now dying) and those who have lost the beloved dwell – whatever or whosoever that beloved might be. For it is in the learnings of those who have lived and traveled in this territory that the way into and through Earth grief is to be found.

Over the course of life, all of us develop the skills of living. We learn to live in *forward time*. Unlike the old and the terminally ill (who are forced by circumstance to give up such an orientation), the majority of people in the world live always with one foot in the future. We make plans for this and that and the other thing, each plan pointing to some future way of being that exists out there somewhere in front of us. And as life goes on, we learn to adapt to the problems that arise, those that affect a forward-focused life. Over the years, as everyone does, when faced with obstacles, we find work-arounds, innovations, alternate pathways that allow our goals to be met, whatever they might be. We develop a "repertoire of problem solutions." The more experience of life there is, and the more encounters we have with novel problems, the more expansive that repertoire becomes.

Eventually most people settle into a more mature, structured life, one in which novel circumstances infrequently occur. The repertoire they have developed is usually sufficient to deal with whatever problems they encounter.

But with a terminal diagnosis or the death of a loved one – when I, or anyone, is told that the cancer or the heart disease or the lung infection is untreatable – that I/we/they are dying – the future is no longer an opening into endless possibilities. We are faced with *ending*. It may come in six months or in two years, perhaps in four if we are lucky. We face the most novel problem a human being can have and suddenly we find that our repertoire of problem solutions is useless.

Nevertheless, each and every human being that is reaches into their bag of this and that and the other thing and brings out solutions that have worked in the past, one by one by one. The solutions are then forced into service with the hope/belief/faith they will effectively deal with this new circumstance. But they fail, each and every time, as they inevitably must. And I suspect that everyone that is and has ever been finds this a terrifying thing, to know that there is nothing that can be done to stop what is coming. That nothing that has been learned over the course of a life is of use now. That what is now being faced is an irrevocable limit. There is a helplessness to it that is terrible to bear and a terrible, terrible sense of a coming *end* that cannot be escaped.

All of us fight against the coming of that ending – either with dignity or without, with honor or without, with bravery or without, with blame or without – for it is in all life to fight against the coming of that darkness. But as the days and weeks and months or perhaps years go by, the truth slowly sinks in. I am dying – my beloved is dying, the Earth climate we have always known is dying, and the life I/we once knew is ending. I/we no longer have a future the way I/we so recently did. I/we no longer exist in forward time. Soon there will be a world without me or my beloved or this Earth climate in it. The life I/we/all-of-us have been living is over.

Everyone faced with a terminal diagnosis (either for themselves or their loved ones – and Earth and its landscapes and many species are very definitely our loved ones) usually responds in a predictable number of ways. (There has been a lot written about this and I am sure you have heard or read some of it.) And of course, somewhere during all this, every denial mechanism that is comes into play. The second, third, and fourth opinions, the rage, the blame, the refusal to see and accept, the immersion in drinking or television or sleep, every one of them.

At their core, each and every one of those responses is an attempt to reassert control, to find a solution so that I/we/they can continue on with life as it was. This is not a bad thing, it's just the way it is. It's just the first part of the necessary process of coming to terms with things. Sooner or later, though, the fact that we or our loved ones are terminal has to be faced. Sooner or later it has to be accepted as an irrevocable fact that is not going to change.

This is the moment we are face to face with the truth that there is an integral-to-this-world ecological limit. The moment we experience that there is no escape from that limit despite all the I-can-lower-my-body-temperature and live-to-be-150 techno-utopianism that is published in the media every single day of our lives. Death is not a flaw in the system, *it is a feature*. It is built into the system and it's built into the system for a reason. Sooner or later it comes for every one of us – just as it comes in time for nations and civilizations and Earth climates, too. And despite our responses, sooner or later, no matter how much we psychologically struggle with it, we will be forced to accept its inevitability – just as every human being who has ever lived before us has had to do – whether we want to or not.

What is most important to the resolution of the journey through ecological loss is what happens at the moment of *acceptance*, when the diagnosis is genuinely received, when it is no longer avoided.

At the moment of true acceptance an important shift occurs. The daily relationship to the self and the world around the self is irretrievably altered. Life is no longer oriented around living but around dying, with the ending of the life that has been led up to this moment in time, with the ending of everything we have known. At that exact moment, the entire structure of the life being lived irrevocably changes. *Forward time*, in the way that it has been known for all of one's life, ceases. There is no more forward time. We are somewhere else now. In a different kind of time. One that is not often talked about in the west.

Much has been written about the "stages" of dying, including, of course, the final stage called "acceptance." I know something about those stages and the processes that people go through to get there for I began exploring this territory long ago, in my late twenties. And I had some very good teachers. People such as Elizabeth Kubler-Ross and Stephanie Simonton were my guides and they taught me a great many things. But oddly enough, what they taught me was what not what I had imagined I would learn.

Stephanie Simonton and her husband Carl worked with terminal cancer patients, those that had six months or less of life left. They were the ones who pioneered the use of guided imagery in the treatment of disease (though other people popularized it). It was not something they had intended or planned on doing. It just sort of happened, as many of the best things in life do.

One of their first patients, as it turned out, was a hit man for the mafia who, despite their protestations, told them far more of his life than they felt comfortable knowing. When every medical intervention failed to help him the man kept insisting that there must be something else they could do for him. Perhaps it was because of his past or maybe it was just time for something new to come into the world but in any event they felt it would be unwise to tell him there wasn't anything that could be done. So, in desperation, Carl told him about something he had heard of recently, guided imagery. The man thought that, yes, perhaps it could help and so Carl began to teach him about it. He told the man to see, in his imagination, as clearly and fully as he could, his body's white blood cells killing the cancer and then his lymph system and kidneys and intestines flushing it out of his body and to do this every day at least once. Well, the man was very driven, very determined, and he threw himself into it. Strangely, and inexplicably, it worked. His cancer completely disappeared. And the Simontons' medical practice – and their beliefs about what was true in healing and medicine and what was not – completely changed. They began to approach disease and the human body from an entirely different place.

> *I remember Stephanie standing there on the stage in that huge auditorium, the room darkened, the lights focused on her, and all of us in the audience captured by her and her words. I remember that moment as I think I always will. For it was the first time I ever saw/met/felt-the-presence-of a truly humble person. I had never seen such a thing in my life. Yet I recognized it the moment I saw it. She stood straight and tall, that incredibly humble look to her face. She had no desire to present herself as more than she was. It was something, I realized later, that had been burned out of her long ago. She was just here in the present, no place to go but where she was, no person to be other than herself.*

She talked for quite a while about her and Carl's work, then paused a moment and said, "There was this woman I was working with some years ago. I had come to care for her very much. We were working, as we always did then, with guided imagery in the treatment of her cancer for there was nothing else that conventional medicine could do for her. It wasn't going too well, the cancer kept progressing, yet she kept at it."

Here, Stephanie paused a moment, looked down and to her right, then gathered herself, looked up at us again, and continued, "I had to travel out of town for a conference and I stepped into her room to say goodbye before I left. I still remember her lying there. That beautiful face.

"Well, when I returned, I was told she had died and that she had left a note for me. I took it from my secretary and walked to my office so that I would have some privacy as I read it. I opened it and this is what it said: 'Dear Stephanie, I am so very sorry I let you down but please know that everything you did meant so very much to me. Much love,' and then she signed her name.

"I was floored by this. I realized in that moment that I had been trying to keep her alive for me. I had forgotten that it was her journey, her destiny, and not my own. My task had been to be in service to her and that journey, whatever it might be and wherever it might lead. But by making it about me I had taken something from her that was not mine to take. And I suspected, quite rightly as I now know, that she had simply waited for me to leave town so that she could die."

*Stephanie sighed, hunching into herself a bit, and I saw such pain cross her face. Then she straightened once more, looked at us (she **really** looked), and said,* "The hardest patient you will ever have is the one who is exactly like you, in age, in gender, in psychological makeup. You will see all these similarities to yourself in them, whether consciously or not. But differently than this patient, you have, during your life, made the decision to live. Yet here they are, making the decision to die. And it *is as if you yourself are making the decision to die. This will touch on parts of you that are very primal. It will undo you. And so, when you sit by them and see them dying you will feel **compelled** to do everything you can to save them, for you will feel, in the deepest regions of yourself, that if you do not save them that you have abandoned yourself to death, that you yourself will die if you do not intervene. You must resist this, for this is **their** journey and not yours."* Then she paused again, took a deep breath, and continued on with her talk.

Later, when the conference was done, as she was leaving there were a great many people coming up to her and following behind as she walked down the central pathway between our chairs. I was on the aisle and, taking my courage in my hands, I stepped up to her and said, "Will you sign your book for me?" And I held it out to her.

She stopped and looked at me then – I will always remember that. It was as if the crowd of people around her had ceased to be and now there was only she and I, alone in the room. Stephanie looked **into** me then, those incredibly kind and wise eyes, full of experiences that I would not myself have for decades, and she smiled this gentle kind of smile and said, "What is your name?"

I told her and she said, "Stephen," and gave a little laugh. "That is my name as well. My parents always wanted a boy." Then she signed the book and handed it back to me, looked deeply into me once more, turned and was gone.

There are teachers who will come to each of us if we merely open ourselves to the possibility of them finding us. They come from the wildness of the world, bringing with them knowledge of the realms in which they have traveled . . . and the wisdom they have found there. If we open ourselves to them they will touch us with something the soul of us needs in order for us to become who we are meant to become. Their teachings will remain inside us like seeds. Sooner or later those seeds will sprout and grow, often when we need them the most, when life demands more of us than we know how to give.

The teachers we find are in this life – and serve in the way they do – in order to spread the seeds they carry into the world. They pass them into us and we pass them in turn into new generations. It is part of the great relay race of soul that has been going on for as long as human beings have been. Elders teach us how to be human – that's an essential part of their ecological function. They are the old growth of our species; we devalue and abandon them at our peril.

First of all – *and to be very clear about this* – the state of mind and being I am speaking of here, the one that emerges on the other side of the journey into and through severe loss, does not come in a single moment of sudden insight as people so often like to believe. Rather, it emerges out of a *process*. (Elizabeth Kubler-Ross was adamant about this and continually insisted to any who would listen that it is *not* linear.) There is *movement* from one state of being into another. (And this movement is more like living inside a windstorm than one step after another.) It comes slowly, each of the learnings hard-won. There's a lot of going back to earlier states, then continuing on, then going back. *Things have to be reworked . . . over and over again.* The process often seems obsessive to the outside observer, as if nothing is happening. But a great deal is happening. *And the obsessive replaying of it is integral to the resolution that is being sought.*

At the moment of diagnosis (or the moment when the sudden, unexpected death of a loved one occurs), the diagnosed and their loved ones are still immersed in forward time – and, importantly, within all the *unconscious* assumptions and beliefs that are part of that way of being. The terminal diagnosis (or the loss) very specifically and directly confronts those unconscious assumptions. Cracks begin to appear. The personal world that is being lived inside of, that has been lived in for so long, so very stable seeming until that moment, begins to come apart. The habituated life begins to crumble. The ship of the self is no longer safe in a protected harbor, it is now setting sail onto the open sea.

The process is resisted, of course, and strongly so. The habituated world is known, loved, safe. But it is going now. And it will not come back. So . . . slowly . . . very slowly . . . those that the terminal diagnosis (or loss) has touched begin to move, one difficult step after another, out of their habituated way of being. They move into *ending*, into the territory of dying. They begin to move out of forward time. And sooner or later, as the reality of what is coming can no longer be ignored, the walls of denial collapse and the full impact of endings rushes in. That is the moment when everything changes. When the harsh facts of life are completely and totally accepted. It is the moment when the old world is totally and completely gone, the moment that grief in all its purity finally finds us.

Fully felt, unrestrained grief has the same relation to sadness that a typhoon has to a gentle spring rain. Until you experience it, you cannot know how completely and totally it can tear apart the world in which you live – the outward as well as the inward. Its powerful winds careen through the self and the life that has been lived until that moment, shaking everything to pieces.

This wild thing that true, deep grief is cannot be tamed by touch or walled off by words. In its arms we become experiencers of grief, expeditionaries of ending, explorers of loss, engaged witnesses who must – if we are to travel through the territory and find the other side – let grief have its way with us. And grief . . . it is pervasive and insistent and relentless. When it finds us, it enters every part of the self, every aspect of our lives. It fills up the senses. The life that existed before, carefully built throughout the years, shatters into a thousand sharp fragments. We live now amid the ruins that loss has made of us, and we grieve. Every day, we grieve.

Grief travels with the dying – and with the families of the dying and the dead – every step of the way. It is the constant companion of endings. Grief comes not only to the dying but *importantly* to everyone who loses someone or something that has been woven into the deepest regions of the self, into the very structure of one's life. At the moment *ending* is finally accepted as the irrevocable thing it is, what is felt is that a *reality* is ending (or worse, has ended), one that has been taken for granted, that has been believed to be as immutable as stone, that is deeply woven into the self. And that loss, it *changes* things. It alters the relationship to life itself as well as to the life that has been lived up until the moment the loss occurred. It alters mental functioning. It alters personal identity at the deepest levels of the self. It alters everything.

As Colin Parkes so eloquently puts it, such a loss . . .

> . . . *invalidates a multitude of assumptions about the world that, up to that time, have been taken for granted. These affect almost every area of mental functioning – habits of thought which have been built up over many years of interaction, plans and routines that involve the [thing or person lost], hopes or wishes that can no longer be realized. Sooner or later every chain of thought seems to lead to a blank wall; [what has been lost] is everywhere and nowhere. . . . Life seems to have lost all meaning.*

Each of us is born into a scenario, into a way of life that, as we grow, we adapt to. At the deepest level of that scenario, foundational to it, is Earth itself and the climate it has had for so very long. Built on top of that foundation are the cultures and nations and the times we live in. (And while these seem to be a foundational reality, they are not, they are *virtual* – as most human beings are now, terrifyingly, beginning to learn.) Most immediate to us in the scenario in which we are embedded are those we love and who love us and of course our bodies and the life we have crafted as we live. Over time, aspects of all these things find their way into the creation of what Parkes calls "an internal assumptive world," something that Val Plumwood described as our "frameworks of subjectivity" – a structure that sustains the "concept of a continuing narrative self" and that helps "sustain action and purpose" in life.

This internal assumptive world is something we build over time, a model of the world that we create and carry inside ourselves. As Parkes comments, "We rely on the accuracy of these assumptions to maintain our orientation in the world and to control our lives."

Relying on these assumptions becomes a deeply ingrained habit, an automatic process that we never think about. Of course, since this is a model of the world and not the world itself, discrepancies are inevitable and so we are forced to adjust the model every so often, throughout our lives. Normally, especially as time goes by, this is fairly easy. It's only when very deep, very-early-in-life-assumptions are affected that serious problems occur, for the deeper the assumptions are, the more they are tied into primal survival drives and essential self identity. It is very difficult to have deep assumptions confronted – by circumstances or by others.

But with terminal events, many, sometimes most, of those assumptions become obsolete. The old world is gone and it *is not coming back*. For those who are dying the entire assumptive world, which is almost entirely predicated on forward time and outward stable, protective structures, begins to collapse. For those who live on after the death of the beloved, most if not all of that assumptive world collapses as well. (And again, for those who are experiencing the loss of the climate and landscapes and species they have been immersed in, and loved, for most of their lives, *what is continually experienced is both death **and** dying*. Loss is not something that ends.)

The scenario into which we are inextricably embedded, in all its complexity (both the foundational and the virtual), is interwoven with our internal assumptive world at a level far deeper than consciousness or rational thought. It is in the deepest regions of who we are, woven into our primal beginnings, for we began absorbing these things into us at birth. This interwoven gestalt is what we call or think of as "reality." Like the moon and the rain it just is. It's rarely questioned. Our personal identity, what we think of as our self, is heavily dependent upon it. And again, because it is interwoven into our earliest pre-verbal beginnings, it resides within us at a level far deeper than conscious thought. This is crucial to understand: it is embedded in us at a level far deeper than our rational minds can go.

When the internal assumptive world falls apart our sense of self destabilizes; sometimes it's lost entirely. It is for most people a terrifying experience, for what is lost are internal structures that most people do not know they have. And that loss occurs *very* deep in the self. One day, out of the blue it seems, the very structure of what's believed to be reality is lost. And with that loss all sense of what is stable and real is gone.

I know what this is like for I have been in that place, experienced that kind of destructuring, more than once.

The conscious mind feels what is happening but almost always is unable to explain it. And because of this it is terrified. With the loss of the internal assumptive world, the person becomes existentially adrift. They lose their feeling of being surrounded by, embedded within, a stable reality upon which they can depend. And the mind, which is very dependent upon the internal assumptive world, can no longer function as it used to. The person loses their moorings to the world around them, to any sense of a stable reality. It is not an easy place in which to be. It is in fact one of the most difficult experiences a person can have, for they become, whether they wish to or not, travelers into the very fabric of reality itself. And that is a place to which very few people in western cultures travel or understand at all.

A very simple analogy (to get a sense of what I am talking about) is that we become like a person who in adult life loses a leg. But until that moment of loss, our sense of self has been oriented around a deeply integrated, far-deeper-than-rational-thought assumption of two legs. This is why every day, sometimes for years after the loss, we sit up in bed in the morning, step to the floor, and find that the support we unconsciously expect is no longer there. And so, we fall. Day after day after day. And every time we fall the new reality *forces* itself upon us. Our old assumptive world, the one so deeply integrated into our sense of identity, fails to hold true. And in shock we are forced to face, over and over again, our loss and the new reality, the new world, in which we live.

Every time we stand, expecting to be supported on that leg, and are not, the fabric of our assumptive world literally tears apart or as Parkes puts it, "There is a rent in the fabric of reality."

Val Plumwood has described the experience of this rent so very well . . .

> *In its final, frantic attempts to protect itself from the knowledge that threatens the narrative framework, the mind can instantaneously fabricate terminal doubt of extravagant proportions: This is not really happening. This is a nightmare from which I will soon awake.*

But we do not awaken. This new reality *is* the waking world in which we now live.

That moment of reality intrusion, what Parkes calls "a rent in the fabric of reality," is a difficult thing to experience. For it is, in actuality, *a rent*, a *tear* in something that was once experienced at a deep primal level to be foundational, so foundational it has never ever been questioned or felt to be transitory.

What was thought of as reality loses its fundamental nature and structure; it is no longer something that can be relied upon. And the shock goes deep into and through the whole system. The impact is difficult to deal with because it is not just the mind that experiences the rent, the deepest parts of the self do. And those parts *believe* in the old identity, in the old world that has been lived in for so long. To them it *is* foundational, it just *is* reality. And to make it more difficult, there are many parts to the deep self – for we are not a single consciousness but multiple personalities existing in precarious balance throughout our lives. Many of the deep parts of the self refuse to accept what is happening, they keep wishing the old world, and identity, back into being. The internal world is disturbed at its most basic levels. There is dismay, confusion, panic.

Every time we stand and find we do not have two legs, we once more experience that rent in the fabric of our reality. We are *forced* to experience it, forced to struggle with the meaning of what is happening. And it takes time, time for personal identity to change, for it to alter itself at those deep, nonrational levels in the core of the self.

The internal assumptive world that you have for so long lived inside of took years to form. It is not so surprising then that it takes years to come to terms with the fact that the form it once had is gone.

With this kind of loss, the real world, the immutable reality that is foundational rather than virtual, the one that exists outside of and beyond our internal assumptive world, keeps forcing itself upon us until, kicking and screaming (though now with only one leg), we finally acquiesce and accept that the life we had so long lived is over and that *this* one-leggedness is the new life in which we now forever live.

And to be clear, the more closely *two legs* is connected to self identity, say if we are an olympic athlete, perhaps a runner, the harder it is to come to terms with the loss, the more existentially bereft we will be. Dick Francis understood this. He wrote a series of mystery novels about a champion jockey named Sid Halley who had lost a hand. In the prologue to *Whip Hand* he captures in exquisite detail what that kind of loss feels like to someone for whom two hands is identity itself.

> *I dreamed I was riding in a race.*
>
> *Nothing odd in that. I'd ridden in thousands.*
>
> *There were fences to jump. There were horses, and jockeys in a rainbow of colors, and miles of green grass. There were massed banks of people, with pink oval faces, indistinguishable pink blobs from where I crouched in the stirrups, galloping past, straining with speed.*
>
> *Their mouths were open, and although I could hear no sound, I knew they were shouting.*
>
> *Shouting my name, to make me win.*
>
> *Winning was all. Winning was my function. What I was there for. What I wanted. What I was born for.*
>
> *In the dream, I won the race. The shouting turned to cheering, and the cheering lifted me up on its wings, like a wave. But the winning was all; not the cheering.*
>
> *I woke in the dark, as I often did, at four in the morning.*
>
> *There was silence. No cheering. Just silence.*
>
> *I could feel the way I'd moved with the horse, the ripple of muscle through both the striving bodies, uniting in one. I could still feel the irons round my feet, the calves of my legs gripping, the balance, the nearness to my head of the stretching brown neck, the mane blowing in my mouth, my hands on the reins.*
>
> *There came, at that point, the second awakening. The real one. The moment in which I first moved, and opened my eyes, and remembered that I wouldn't ride in any more races, ever. The wrench of loss came again as a fresh grief. The dream was a dream for whole men.*
>
> *I dreamed it quite often.*

It is so very hard to let go of the old self (as I have found from painful experience), of who I once was – that marvelous, healthy, whole and strong person that I took for granted and believed would last forever. The one I just accepted at a level far deeper than words can ever go. It was a life-spring and I lived and saw the world through its eyes, did not know that any other way of being could actually exist. Then, one day, it was gone and no matter what I did or tried, it was not going to come back. And so, I dreamed. Of who I was, of what I was, of how I was. Night after night after night. It has taken so very long to find acceptance.

I think that all of us who enter the world of terminal illness, of the *loss* of what we were, go through this.

But at *acceptance,* the strange country that takes so very long to find, all that changes. I am no longer the person I was, no longer wishing the old world into being. I no longer grieve the loss so keenly. And with true acceptance I was no longer bitter, not angry, no longer blaming, no longer living in "if I had only" or "if they had only." Those ways of thinking have disappeared. I am someone else now. Somewhere else now. My entire orientation and relationship to life and self and culture has shifted.

That is what it is like with something as life changing but as simplistic as losing a leg. And a leg, for most people, is far easier to deal with than the loss of a beloved child or spouse or to be told that we, ourselves, are dying. It is easier to deal with because you can more easily see and feel that it is gone. It is a simple physical fact. But with the loss of a child or a spouse or our own forward-living life, it is much harder. There is no leg to point to that reminds you every day of the fact of what has happened. (Your beloved might, after all, just be in the other room, as they so often have been before.) And this makes a very large difference in how you deal with it. Simple physical reductionism doesn't work so well. Something far more invisible to the eye and the reductive mind is in play.

Those we love and who love us (just as Earth itself is) are woven throughout our sense of self, into our thoughts, our plans, our days, our hopes, our futures. They are the source of the deepest intimacy we will ever know, the ones who companion us and whom we have come to trust with our most vulnerable self, the part or parts of us that no one else is ever allowed to see. A million times every day we reach out with some invisible part of us and touch the living reality of them, just as they do us. And that touching, that companionship, that trust in an outside someone who loves and believes in us, is intricately interwoven into our sense of self and our relation to the world around us. They are the existential legs upon which we stand, that support us, that we rely on at the very deepest levels of our being. When they are taken from us that invisible, woven-deep-into-us support is lost. We have nothing to stand upon at all. And we fall in the most terrible sense of falling. Over and over again. We have lost the ground of being around and upon which we have interwoven our self and life. And that loss brings with it a terrible alteration of our internal assumptive world. As Parkes observes . . .

> *We can only recognize the world that we meet and behave appropriately within because we have formed models to interpret our perceptions and guide our behaviors. We recognize chairs, tables, doors, and windows because our internal world contains memories of all these things on the basis of which we make reliable assumptions about them. We walk through doors with confidence because we have learned at the deepest levels of our mental processes that doors set off one region of solid footing from another region of solid footing. . . . it is unlikely that we shall meet a door that looks like any other door but leads into an elevator shaft or empty space where a now-demolished floor once existed. [But after this kind of loss we continually encounter] empty space where security once was.*

After such a loss, every moment, every breath, every thought finds us stepping through a door expecting to find solid ground. But we don't. Instead all we find is an empty elevator shaft. Every morning we wake up and unconsciously stand, expecting solidity, but it is no longer there. We fall and the fall is endless. There *literally* is a "rent in the fabric of reality," one that tears itself into and through every part of the self. And we feel it every moment of our daily life. The old world that we relied upon for so long is gone and that reality, that world, will *never* return. We have not just lost a leg, we have lost the kind of companionship that is very hard to find in this life. We have lost an integral aspect of our identity, a mirror that has told us for decades who and what we are. We are now existentially bereft.

In that moment, we find that our sense of self is gone. And that sense of self has come in large part from innumerable intangible structures embedded within us, which were in fact a kind of psychological/spiritual skeletal structure that was holding us up and giving us our shape. Now, it is gone. We can no longer *feel* the structure of reality around us. It has simply ceased to be.

Afterward, we look out into the world and see, as we always have, that yes that is a chair, that is a door, that is food, that is the outside world, that is the sun and green grass and children running in the field. But there is no longer a personal connection to those realities, they have become only intellectual facts. The *meaning* of them is gone. And that takes us into a very particular and peculiar world, one that can only be found/experienced when foundational assumptive structures are completely and totally lost. Again, as Denise Riley describes it . . .

> *Wandering around in an empty plain, as if an enormous drained landscape lying behind your eyes had turned itself outward. Or you find yourself camped on a threshold between inside and out. The slight contact of your senses with the outer world, your interior only thinly separated from it, like a membrane resonating on a verge between silence and noise. If it were to tear through, there's so little behind your skin that you would fall out towards that side of sheer exteriority. Far from taking refuge deeply inside yourself, there is no longer any inside, and you have become only outward. As a friend, who'd survived the suicide of the person closest to her, says: "I was my two eyes set burning in my skull. Behind them was only vacancy."*

The former, comfortable, reliable world and the sense of self, of personal identity, that emerged from of it are gone. And far too often, people are so terrified of this *feeling* of non-meaning, of existential emptiness, that they flee to their physicians or their psychiatrists and are given medications so they won't have to face it, won't have to face the reality that now exists inside them.

I know the terror of that place for, as I have said, I have been there more than once. And I have found over the years of my life that the only true solution to it is to one day decide not to flee but to turn the face toward it, to enter its emptiness, to descend and discover what it is trying to teach.

> *That feeling of non-meaning, it is important to understand, is **only** a feeling. And the fear, the terror? They are **only** feelings as well. It is possible to become accustomed to them. To get used to their presence in your life. The truth is that you are just unmoored. It **will** end. There is indeed another shore. And no, I did not believe it the first time I descended into that void either.*

Those who have lost their loved ones are clear about how it feels to them once they have. They speak of it perhaps less eloquently than Riley does but it is no less heart-rending: "My husband's in me, right through and through" and now "I feel as if half of myself is missing" and there is "a great emptiness" inside me. "I had heard that expression 'heartache,' I had read it, but this is the only time I really knew what it feels like. It is pain inside me, physical pain, all the way up. It's very tight and I get very hurt. It's inside the heart that I hurt."

This is what it is like to be in deep grief, to lose the beloved, to lose . . .

> *The flesh of my flesh, bone of my bone, my breath, my heart, my reason for being, my everything. My beloved is woven deep within and throughout me, entangled in my very being. And now she is gone . . . and . . . I am gone.*

The loss of the beloved forces, perhaps for the first time in a person's life, a grappling with fundamental questions of identity, of who we are, what we are, even the purpose of our individual life.

Long ago, Viktor Frankl spoke of this sudden casting of the self into loss of meaning. My memories have altered his words over the years that

I have remembered them. Here is the shape they take inside me now . . .

> *There comes a time in every person's life when, in distress, we leave the house and in darkness walk to the top of a small rise, look up at the stars, and say,"god, what is the purpose of my life?" We do so without realizing that we are not the questioner but the questioned.*

Whether we wish it or no, significant loss shatters the old internal assumptive world. We look out at the world and nothing is as it was. There was a vital, living world around us, now there is not, only the intellectual fact of a vital living world – and that is something else entirely. We see the world but we have no emotional connection to it, no feeling of meaning, no response of the heart to its existence. There is only the intellectual fact of it.

And so we are asked a question – though at the time what is more accurate to the experience is that we feel that a question is being forced upon us. And while it is a single question, it has many parts: *Who am I? What am I? What do I do with my life now? Where do I go from here? What is the purpose of my life? What is the meaning of life – of my life? Why go on at all?* This is *in reality* the most important question any human being can be asked. How that question is answered determines the shape of the life we live afterward – for all of us live two lives, the one we learn with and the one we live with after.

There are, roughly, three aspects to the journey that occurs after this kind of loss. The easiest part of the process (easy only in comparison with the other two) is accepting the loss intellectually. It's the simplest aspect of the new reality: a person was here, now they are not. (Earth was this way, now it is not.) We can literally see the absence. (The forest we loved is gone, it will not come back.) We wake up in the morning, turn over to touch our beloved, and they are gone. We wake up in the internal assumptive world we have always known (it has been with us so long that it asserts itself automatically, every day of our life); it wraps itself around us, comfortable, safe, and warm. But every morning, when consciousness fully returns, when *the second awakening takes place*, that habitual world is confronted with the new reality. And one way or another, as the days progress, intellectual acceptance is forced on us. No matter what room we look in, the beloved is not there. What we loved is gone and it is not coming back.

But accepting it emotionally? That is another thing altogether. And the final step, constructing a new life? That comes as time progresses; it's inextricably interwoven with the emotional acceptance of the loss.

"Emotional acceptance of the loss." It sounds rather simple, doesn't it? Most people interpret this as "getting used to it." But there is a great deal more to it than that. And it's not easy because what is being grappled with is the loss of the internal assumptive world. The loss of the very structure of reality itself. It is a loss that is constantly demanding an answer be found to the question "Who am I now?" It is not an easy question to answer. And all the time that the answer is being sought one lives in the world of grief and loss and emptiness. In the midst of hopelessness and terrible, terrible emotional pain. In a nebulous state where *meaning* in its deepest, existential sense is gone.

In a very simplistic sense, the internal assumptive world can be thought of as a set of operating instructions, software, that has been assembled throughout the course of a life. But, again, it operates at a level deeper than rational thought or consciousness normally goes. It serves a great number of important functions – most important of course is the unquestioned, constant *feel* of reality around you. There is a reflection that comes from it, as with a mirror; it tells us who we are, that is, it gives us a sense of identity. Two other very important aspects to this unconscious, unquestioned sense of meaning are: shortcuts and a feeling of safety.

Shortcuts mean that we don't have to, each and every time, analyze the environment in which we find ourselves, then laboriously craft an interaction in response to it. We recognize patterns and implement solutions. The recognition and implementation process happens very quickly, in milliseconds once the internal assumptive world is constructed and has matured with experience. (Small children, early in their life, and adolescents – that is, when our child world changes as we become sexual beings – have to spend a lot of time figuring things out.) Shortcuts allow us to just get on with life and not have to figure stuff out all the time.

> *Though maybe, importantly, the shadow world that most adolescents live within for so very long, and which most people take for granted, is in actuality grief. The grief of losing their younger, child self and way of being, accompanied by the lack of information about what is happening to them now, their lack of a map, the lack of honest conversation about a common shared journey all of us take. It is no wonder then that so many of them are sullen, that when asked what is wrong they remain mute, perhaps cross their arms across their chest, or gesture impotently to the region of their chest, the place their heart resides.*

Safety means that we know how to act in the situations in which we find ourselves; it is the feeling of competence within a known world. And most people after adolescence, it is important to understand, stay within a particular set of parameters, ones they know well and are comfortable within, Because of this they feel safe. They are not explorers, immersing themselves into new territories that are strange to them. They are not psychonauts, exploring new regions of the psyche, dismantling and recreating in new form their internal assumptive world as they travel into new geographies of mind and being. Instead, most people find their niche and settle into it. Every day, upon waking up, they put on their life again, that familiar internal assumptive world, like putting on a pair of comfortable old shoes. They are surrounded by the known. They feel safe. And this is not a bad thing. It is what most of us do.

As they move through the world, even though they don't know they are doing it, they use evolutionarily-developed senses to check for anomalies, disconnects between the assumptive world and the real world. Most people, over time, find ways to avoid anomalies. (Homeless people, unsafe neighborhoods, the working class, and yes I am being ironic, which comes from the ancient greek word *eironeia,* meaning "simulated ignorance.") They find their bubble and stay in it. (*Everyone does this* and no there are no exceptions, there are just different kinds of bubbles.) And while what each person finds disturbing to their internal assumptive world differs, everyone finds ways to avoid what disturbs it.

When the internal assumptive world collapses we lose the sense of meaning that we have had. We lose a sense of self. But, as well, the world no longer feels safe. Suddenly we are in a world that is strange to us, one for which we have developed no skills, in which we have no experience. *Nothing* is the same anymore. It's unsettling in the deepest sense of that word. As Parkes puts it, the bond with the beloved created . . .

> *. . . feelings of being somehow secure, augmented, extended, or completed by another which make it possible, when the marital partner is present or at least accessible, to be comfortable, relaxed, and so able to give attention to other matters. [The] loss of an attachment figure . . . means loss of a critical security-fostering figure. It brings about a sense of being alone, beleaguered, vulnerable. . . . [T]hose who are confronted with a sudden disaster tend to turn for help to the people to whom they are attached. But [with the loss of a loved one] the person to whom the individual would normally turn is the very person who has been lost. And so the anxiety continues unabated. . . . Faced with awareness of a sudden massive gap between the world that is and the world as it should be, and with the sudden loss of the security-fostering figure, [those experiencing the loss are] required to deal with a truly overwhelming threat.*

With the loss of the internal assumptive world that feeling of safety is lost. With this comes "great restlessness, difficulty in concentration, difficulty sleeping, anxiety, and tension . . . intense sorrow, painful memory, hopeless pining for the lost figure." And of course, depression.

> *Depression is the way the world makes what we've been hiding in darkness visible. It is a communication to us that an assumption we have is inaccurate to the reality of ourselves or the*

world itself. It stops forward motion – which is part of its essential function. It's a message from the depths of us or from the world around us – sometimes both – that there is an unexamined assumption that is in our way. That it must be examined and changed. The deeper and more primal the assumption, the greater the depression. The longer we refuse to examine and then change it, the more intense the depression becomes. The depression almost always involves self identity or plans we have (sometimes totally unconscious plans) that have not included something essential to who we are or to what the real world is or both. With the loss of the beloved, the assumption – that is, the internal assumptive world – built around the beloved is no longer accurate. That life, that assumptive life has to be carefully sifted, examined, then changed.

It is always helpful to remember that depression is an integral part of loss, of grieving, of the journey to the other side. It's not a bad thing, even though it feels so terribly frightening. It's just a communication, just a feeling (as overwhelming as it might feel). It will pass. It is just part of the territory.

All these feelings and states are a natural response to the fundamental disruption of the world that we have constructed from our bond and that had become for us an unquestioning reality.

When the *meaning* of things is lost, there is *always* the loss of the formerly unquestioned internal sense of safety. We become uncertain. Assertive action is inhibited. We no longer know *how* to move through the world; we have lost trust in the stability of life. It is no longer dependable. The world we are in now is *very* strange and quite often forbidding in how it feels to the heart. It is not the comfortable world that we once knew.

For many people the loss of trust is so extreme that anything more complex than sitting in the intellectually-recognized chair in front of them is very difficult. With every movement they make (no matter what it is) they find only an elevator shaft, not solid ground. In consequence, many people withdraw into themselves and, if they can, stop doing much at all . . . at least to an outward observer.

But in their internal world a great deal is going on. Every person, no matter who or what they are, is *replaying*, over and over again, the events of the loss itself – and the feelings that go with it. There is a reason for this, an important one. As Parkes says . . .

> *The process is difficult, time-consuming, and painful. It seems that emotional acceptance can be achieved only as a consequence of fine-grained, almost filigree work with memory. It requires what appears to an observer to be a kind of obsessive review in which the widow or widower goes over and over the same thoughts and memories. [However] if the process is going well, they are not quite the same thoughts and memories, there is movement – perhaps slow – from one emphasis to another, from one focus to another.*

Grief *demands* an obsessive review of past events, over and over and over again, always to the point where friends and family just want to scream. What outside observers don't understand is that this obsessive review is essential to both emotional acceptance and the rebuilding of the self. And, of necessity, it occurs one incredibly slow step at a time.

> *And now the final source of shame and guilt reveals itself. Shame and guilt, in some form, always come with the death of a loved one (including that of Earth ecosystems and forests). When a loved one dies it is common to feel that we have in some way been at fault, that we have contributed in some way, even if we have not. We were not aware enough, we didn't cherish them enough, we didn't insist enough that the doctors look deeper. It is this sense of not paying attention that is perhaps hardest to bear. For the inevitable belief is that if we had been paying attention things might have been different – they might have lived.*
>
> *As well, not only were we not paying attention, we were off in our own little world having fun, worrying about inconsequential things, while the one we loved was getting sick. We*

> *didn't even notice. Rightly or wrongly, there is a feeling of being at our core a selfish, unthinking, blind, and shameful person. And too, inevitably, we begin to remember all the times we were unkind, when we weren't attentive enough, start thinking of all the things we should have said but did not. And so . . . on and on and on. Some people call this "survivor's guilt" but I find that too facile. In truth it is an essential part of the reordering of our internal world and our relation to the outward. It is an integral part of coming to terms with the loss.*

Everything has to be replayed, over and over again. Part of what is being analyzed, despite its appearance as merely an obsessive replaying of pain, is the degree of personal responsibility for what has happened. And again, while the replay appears to be identical every time, it isn't. There are very subtle, tiny shifts, what Parkes describes as "fine-grained, almost filigree work with memory." (What a beautiful line that is.)

> *Patton Oswalt said that after the sudden death of his wife he still had his daughter to care for so he **had** to get up every morning and do the work of being a father whether he wanted to or not, whether he felt he could or not. He found, over time, that doing the laundry, washing the dishes, vacuuming the floors, odd as it might sound, was integral to the rebuilding of himself. Somehow, as he did those mindless chores while obsessively thinking of his wife, his soul-shattering loss, his failures to see, his emptiness, his internal world was being rebuilt in a new form. And deep within his insight is an odd yet important truth: Doing the laundry is in some inexplicable way cleaning our interior world. Vacuuming the floor is somehow clearing our interior of accumulated dust. Making food for our child is learning to engage in the sacred act of breaking bread with our self, this new self who is coming into being now. It is also learning how to, once more, engage in the sacred act of breaking bread with the world again, but with this new self, the one that has oh so slowly come into being over these past days and months and years.*

The review process lasts as long as it lasts; there is no "healthy and normal" period of time to it. (As my grandmother once put it, "My mind just wasn't right for five years" after her husband died.) Irrespective of what the outside world thinks, rebuilding the self takes time and patience and the slow work of years. But, commonly, as time progresses, there is an increasing insistence by outside observers that the review process be terminated sooner than it needs to be. And this can come as well, though it should not, from psychotherapists of all persuasions – long-lasting grief, more than six months or so, is quite often considered to be pathological, as so many other of our internal states regrettably now are.

There will be increasing pressure on the grieving person to take pharmaceuticals to short-circuit the grieving process or to "buck up" or to "get over it" or to "get out and do something" or to "spend time with friends." As Parkes comments . . .

> *The repeated review by which emotional acceptance is obtained can be painful to friends and relatives, as well as to the widows and widowers themselves. Friends and relatives may urge that the review be terminated long before the widow or widower has adequately come to terms with the past.*

In nearly all instances, this is simply because the outside world is so terribly bad at dealing with pain – of any sort, physical, emotional, or that which comes from soul damage. It is *very* hard to be in the presence of another's excruciating pain month after month after month, year after year after year. If the process is not terminated, then one by one friends and most relatives will step away, go back to their lives, to once more living in forward time. And as Parkes says, "What this can mean is that after a time – often, a rather brief time – the widow or widower is left alone with the work of review."

> *Much of your life now is spent in the place that Parkes calls "the loss of the internal assumptive world." It is not something that just comes and then goes. It is, for a long while, a habitation of the self.*

> The long engagement with minutiae, the excruciating and exacting review that occurs while you are in that habitat, is important. It is integral to the working through, at the minutest levels of your being, of the loss of the old world – which includes your previous self identity – and the subsequent, slow crafting of a new self and world, one that rests on a sounder foundation. The one that Greta was speaking of when she said that false hopes are not the way.
>
> It is crucial to understand that you who are coming to terms with ecological loss and what it means are in the process of accepting and integrating something that our entire culture does not want to admit of or face. It is rare for any of us to find companionship on that journey. Because once you accept the diagnosis, the entire framework of what you call and feel to be reality shifts; it shifts substantially. You are now outside the frame that you once had and that the culture and the people around you still have. They are in one world, you are in another. And while you can understand the world they are in, they cannot understand yours.
>
> This means that, of necessity, you will have to review every relationship you have with the people around you. And you may find that you don't have a lot in common with them any longer, nor with the culture in which you have been so long immersed. You really are someplace else now.

Very few people will enter this place of loss and review with you, say nothing, just witness what is happening – holding you when you need to be held, leaving you alone when you need to be in solitude. Very few people can understand that you are on a journey, a journey filled with loss and grieving and . . . eventually . . . renewal – and that the only thing they can do is companion you as you travel.

Those who understand this kind of companioning are physicians in the real sense of that word (whether they have an MD or not), that is, "those who work to alleviate suffering." The point is not "cure" (for this is not a disease). A focus on a "cure" for our grieving, as Elizabeth Kubler-

Ross has said, is all too often a cover for denial, for fear, for the terrible anxiety that comes when death or severe loss threatens to enter our lives.

Eventually, for those of us who grieve terrible loss – whatever its nature – there comes a moment that appears to an outside observer as (legitimate) *movement*, that is, the first careful steps outside the house, the first tentative engagements with the outside world again. It is very similar in its nature to a process that Gary Snyder once described about the writing of poetry. Here is Robert Bly's description of it, from his piece, *Hearing Gary Snyder Read*.

> *He speaks softly before the student audience, confident that he has much to say, and it is exactly what they need to know. He makes a few remarks about [his poem] Rip Rap to start with. On certain mountainsides in the far west where one might want to build trails, an obsidian rock sheath is found, glassy, impossible for horses' hoofs to get a grip on. So smaller rocks have to be laid on it, but carefully. So he thought that words might be used that way, one slipped under the end of another, laid down on the glassy surface of some insight that one couldn't stand on otherwise.*

That's a beautiful metaphor, isn't it? And it perfectly captures what the slow movement back into the outward world feels like. People are still in that existential state where the meaning of things, the old, pre-verbal beliefs and assumptions about life, are gone. They are in that particular and peculiar habitation of self. But from of all the obsessive replaying some progress, however tiny, has been made into emotional acceptance. Into an explanation, however tiny or incomplete, of **why**. And with this comes a tentative restructuring, deep in the core, of self identity.

Why is a question that always must be answered in a way that both the soul and the heart can understand – it is not and never has been a mental answer that is needed. And it has to, as well, be an answer that makes sense to the four-year-old child that still lives inside each and every one of us. This means that the normal answers that people give, which all too often come from the dissociated intellect, will never do. With deep loss we are forced, always, to grapple with why. And *why* is a question that is very deep indeed.

So, too are the other questions that come: Who am I now? What am I now? What is my life going to be now? Why go on if this is the way it is? And over time, during all that endless internal work, initial answers of one sort or another come. They are always, in the beginning, tiny, wobbly, uncertain answers. They are *suppositional*. In aggregate these are the beginnings of the new meaning that the core of the self is seeking.

Nevertheless, these answers are tentative, only the beginnings of a trail that can be laid across that glassy, slick mountainside. But maybe, we think, they are sound. So we lay the first rock down, then another on top of that, and then another. And we step out, hoping for solidity. But the new trail doesn't hold. We fall. And the fall down that mountainside of slippery, sharp stones is painful and terrible indeed. So, another retreat into the inward occurs. More obsessive reworking takes place, then another perhaps more stable understanding comes, a more stable meaning is created. Once again a rock is laid down, then another on top of it, and another. Once again we step into the outward. And once again we fall. Sometimes it takes years to find the stability that is sought. But all that falling, all that reworking, all that contemplation, all that sitting in non-meaning – we learn from it – painfully we learn. And as time goes by we build a new self and a new life one piece at a time, one meaning laid down, then another on top, and then another. We rebuild our internal assumptive world, one far stronger and more adaptable than before.

More than anything, this is an *experiential* process and we can only find out how stable the meanings are when we lay them down across that slick mountainside and begin to walk upon the trail they make. As Parkes says, "[The process] is tenuous, difficult to maintain, and easy to interrupt." But we learn, and we grow, and the new self gradually comes into being. And when it does, we see the world in a very different way than we once did. For there is one thing we now know at the deepest levels of our being: safety does not exist as a permanent state; it can never be foundational to life; trying to make it so is a fool's errand. We know at the deepest levels of being that *change* is the foundational truth here. That fundamental, foundational *endings* exist. But more than that ...

Grief is now interwoven deeply within us. Existential loss is woven deeply within us. And from this interweaving comes *gravitas*, a centering, an existential stability which to most people is a new thing. It's as if we are a ship and during our long grieving we have constructed a keel that now extends downward into those dark depths of loss in which we have lived for so long. And the one thing that keels do is they keep the boat of our life balanced and stable as we move across the surface of the sea. And in the process, inevitably, we learn the difference between optimism (which is always concerned with safety and "positive" outcomes) and hope (which is not).

There is a huge difference between those two states. Those who are optimistic, that is, those who wish for a return to a state of innocence where there is no pain or conflict or loss of safety, do not possess a keel of any depth. The keel which keeps our ship afloat no matter what terrible storms afflict us always extends itself into the darkness of severe loss and grief, into that *truth* of life. And during the slow creation of that keel, unexpectedly so, a certain kind of faith arises. It's called hope.

Hope is faith in life itself. Not in human life, but *life*. All human life (ours included), all human creations, end. Even our civilizations end. Even the climate of Earth ends, as it has ended so many times before. But life . . . that is inherent in Earth itself. It is something that continually extends itself upward out of this living planet we love. It is the substance we see in the face of the puppy that loves us, hear in the squeal of a child's laughter, feel in the emerging of green sprouts in the spring. Life.

When we begin to move outward into the world again, we feel that substance once more, that living field of energy that we were once immersed in all unknowing. This new heart that has come into being feels the life force again as if it were the first day of life. We feel the faith that life is. And out of nowhere, unlooked for, we find our smile again. But it is a smile tinged with a sadness that will never leave, for grief has touched us in places that will never be free of that touch. Its tears have carved arroyos in the landscape of our face, our body, and our heart. They will never leave us in this lifetime. Those who know how will see what we now carry in our eyes. For our eyes have become far deeper than they once were; some shadowed pool resides within them now. In that pool are the teachings that can only be found during loss and the weeping that grief demands of us.

And the movement outward across that slippery slope? Those falls, the retreat inward, over and then the movement outward again? It is the oscillation that occurs between death and life, back and forth, and which moves faster and faster as time progresses, until *death/life* integrates into one whole state of being, where death is understood to be inherent in life and life inherent in death. There is no longer death *and* life but *life/death*. What has been found then a place beyond *either/or*. It is a place where a new kind of understanding and truth emerges at the core of personal life, the first instant of stability on the other shore, on the other side of grief. It is the emergence of the keel, the emergence of *gravitas*, the beginning of the new life.

I cannot tell you, as no one can, what the new self you find will be like. For each and every one of us who enters that territory is answering for ourself the most important question there is, *Who am I?* And the answer to that question is always unique, individual, and crafted over slow time out of the deepest regions of the self. It comes out of what truly matters to you.

You should know that, regrettably, close friends, family, and coworkers may not like or even be able to relate to the new self that is coming into the world. The foundation of personal relationship to the outward has been completely altered. More than most people realize, that old internal assumptive world was also deeply interwoven into every human relationship that had formed. It provided a common framework, a common understanding of reality. Once you change that common understanding, relationships, quite often, do not survive.

The slow construction of a new life, that is, the decisions about how you will be, what work you will do, how you will approach the world and yourself, occurs concurrently with the emotional coming to terms with the loss. The pain that has been so much a part of life, so much like a broken tooth that the tongue returns to over and over again, or perhaps more properly, the heart returning to what has been broken inside, feeling it over and over again, changes. Somehow, because of all that continual touching, all that deep work over the long years, the sharp edges have been worn smooth. They are far less sharp and cutting. And now the pain can be carried, just part of the weight that life demands of each of us.

You are in the new life now; you are not the person you were.

INTERLUDE FIVE

"It's Hard to Live without Love, You Know"

Foreknowledge [that the condition is terminal] is important to recovery, but it does not greatly reduce the need to grieve.
COLIN PARKES

I have always felt that a human being could only be saved by another human being. I am aware that we do not save each other very often. But I am also aware that we save each other some of the time.
JAMES BALDWIN

Oh, I see, you were being a bridge. That's nice. But you see, when you're being a bridge, you yourself never get to cross over.
NAN DEGROVE

Elizabeth Kubler-Ross sat in front of the room – in a chair as she always did – smoking and outraging purists of all kinds as she watched the room with those clever eyes of hers. There was a bunch of us in a semi-circle in front of her, sitting on a scattering of pillows on the floor. Others were farther back, seated among the rows of chairs that filled the back half of the room.

I'd never met anyone with less sense of style than Elizabeth. She seemed to have no time, or tolerance, for those sorts of things. Things like hair (she cut it short as if saying "There, that's taken care of. Now, let's get on with things."), and makeup (What? This is the face god gave me, why change it?), and those clothes (They keep the rain off and take care of your social conventions, so, done). She simply *was*, perhaps more than any person I've ever met. Yet I'd noticed, very early during the days I spent with her, that she was always acutely aware of what was going on around her. Her eyes didn't just look, they *saw*. And when any of us spoke, she really *listened*. And her mind was constantly processing everything that she saw and heard.

I have come to think, from all these years of contemplating my time with her, that it wasn't merely that Elizabeth had no time for feminine pursuits (as people sometimes called such things then) but that she had intentionally adopted a very-carefully-designed camouflage that allowed her brilliant mind and careful observing to go unseen by most outside observers. She had, after all, made waves in what was, when she began, a man's world. And she had been unapologetic about it from the beginning.

For many years, the physicians among whom she worked called her Dr. Death and they didn't mean it as a compliment. Appallingly, she refused to hide from her patients the fact that their illnesses were terminal. (Which most doctors did then and which some still do.) To make it all the worse, she talked out loud about death pretty much all the time, forcing male physicians to face something they had spent decades avoiding.

I found her refreshing, they found her rude and unprofessional.

As usual, the workshop I attended was filled with a large variety of people. Some were dying of cancer, some had lost or were losing loved ones, and there were, as usual, therapists and healers of every sort. The days I spent with her would change my life. . . in more ways than I had ever imagined they could.

She created a place where the importance of a fully lived life and an honest encounter with death at the end of it were spoken of openly. A place where truth was welcome. A place with no evasions or foolish optimisms. A place where people talked of dying and death and what it really meant, where there was no mention of passing away or crossing over. And because of it, as the days went by, we moved from superficialities into intimacy. We began to show our true faces, reveal the hidden dimensions of our hearts. We unconcealed ourselves to the gaze of others who were, simultaneously, unconcealing themselves to us (which is, when you really think of it, what true intimacy is). And as the time progressed, we coalesced into a close, intimate group.

It was about a week or so into the workshop when Elizabeth said . . .

"By the way, because there is a tendency to develop quite a deep intimacy during these workshops, it is common for participants to want to have a reunion a year or so later. I don't believe in them and strongly suggest you don't do it. In fact, in all my years of teaching there was only one time that such a reunion did occur. Here . . . I will tell you about it.

"It all began when I was in Hawaii teaching one of my workshops on death and dying . . .

"That was during the early years of my teaching, just after my book *On Death and Dying* had come out in fact, and my work was not so well known then. So the group was small, perhaps forty people or so. Most of them were sitting on pillows, as so many of you are now. (And here, she gestures with her hand at those of us on the floor.) Though, of course I had my chair to sit in, just as I do now. (And here she pats its arm.)

"We had all gotten our tea and settled ourselves and just as I was about to begin the door at the back of the room opened and this very elderly woman came in. She was dressed very formally, all in black, and had such an expression on her face – as if she had just tasted lemons.

"She strode up to the circle, looked everyone over, stared at me, and then said, 'Well, where's a chair for me. I can't possibly sit on the floor like *that*.' She said it as if we had asked her to live in a homeless encampment near the highway.

"We searched the place and the only other chair we could find was this old, upright, spanish inquisition sort of thing that took three people to carry into the room. They set it right next to mine of course whereupon the woman immediately seated herself and began to glare at the people on the pillows in front of her. Then she pulled out a cigarette, put it into a long, elegant holder, and said, 'Well, who's got a light?'

"As you can see, in this pouch around my neck I carry a lighter with me at all times. I pulled it out, leaned over, and lit her cigarette. The woman sucked in the smoke as if it were life itself and said, 'Well, get on with it then.'

"And so the week began. To start with I asked the people attending to share why they had come. Unsurprisingly, the old woman was the last to speak. The room was silent a bit, the people looking expectantly at her, until finally she sighed deeply with a sort of 'if you must know' kind of sound and said, 'My children paid for it and in my world you don't waste money.' And that was it, the last time she spoke for the entire week.

"As usual the week was filled with incredibly deep and painful stories of loss. There was a great deal of weeping, of unchained grief, of stories told by those with cancer and those who had lost loved ones to it. All the time this was going on the woman and I sat side by side and smoked. She, continually leaning over and asking for a light, me reaching over and giving her one.

"At the end of the workshop, as everyone was gathering their coats and shoes, I took her aside and asked her why, since this was so obviously a place she felt she did not belong, she had stayed for the entire week. She looked at me and said, 'Well, you smoked, so you couldn't be all bad.'

"At that, I took the pouch from around my neck and gave it to her. She gave me a rather sharp glance, then nodded and left.

"It so happened that a year later I was again in Hawaii, for a medical conference this time rather than a workshop. I was contacted by the group from the year before and told that they had, to my dismay, organized a reunion. After much agonizing I decided that this one time I would go.

"It was festive affair. There was alcohol and food and a great deal of happy laughter. About an hour into it, the doorbell rang and into our midst walked a woman that I did not remember from the workshop. She was dressed in flowing purple pants and a crimson, billowy blouse. It was only when I recognized my lighter pouch around her neck that I knew her.

"And oh, but she was the life of the party, laughing and entertaining everyone with her stories. Then, many hours later, she suddenly grabbed her chest and said, 'oh, my!' and fell to the floor. Everyone crowded around as I checked her vitals. Then, carefully, we lifted her and took her into one of the bedrooms, settling her gently on the bed. She motioned me closer and as I bent over she took my hand and whispered to me, 'Thank you for teaching me that it's never too late to change.' And then . . . she died."

It's been forty years since Elizabeth told us that story. I wept when I heard it then and I weep again as I tell it now. For something true and real comes into the room every time I do and it always touches me deeply. Which is why I suppose I have kept it close all these years, carefully wrapped up inside me in the place where I keep all such things.

Every so often I take it out and unwrap it and let it come alive again. When I do, in memory, I see Elizabeth sitting there in her chair, those watchful eyes of hers, taking us in. Her *seeing* going deep inside each and every one of us, reaching places that few of us even knew existed. Then, at just the perfect time, into a moment of silence, she takes those words out and carefully drops them one by one into the room . . . and into us, where they taught us, too, that it's never too late to change.

She told us three stories during those weeks; I have kept them close to me all these years. Each of them changed my life. Shall I tell another of them now?

During her early years as a physician, Elizabeth worked in Denver treating the terminally ill, most of whom were suffering from cancer. She wore those absurd dresses that were popular with women then and those incredibly ugly pointed-on-the-sides glasses that were then in vogue. To make the glasses even more ridiculous women sometimes had rhinestones on the edges, sometimes on the glass itself. The rhinestones might spell out the woman's initials or other times their whole name. (My mother wore them for years; I still have nightmares.) And for some reason women's hairdos were often sculpted into a high bouffant style (it's a french word, meaning *swollen*) that needed a can of hair spray to hold it in place. It crunched when touched (I still have nightmares) and I could not understand why anyone would do such a thing.

It was a strange time and I'm glad it's gone.

Most of Elizabeth's patients didn't have long to live. Over the weeks she worked with them, they declined slowly, inevitably, until one day they began to talk to the nurses or their doctor or their family about the trip they would be taking soon. Or the ball game they had to attend and are you sure you have the tickets? Or I'm going to visit Mother soon, I am so happy, I've missed her. And then a day or two after that they would die.

But one year, Elizabeth noticed something odd. A number of her patients, all nearing death, were getting better. Not going into remission really, just getting *better*. Their spirits were up, their immune function stronger, the cancers slower growing. And she couldn't find a reason for it. Their condition tended to worsen over the weekends, but when monday came around again they started getting better. Then the weekend came and they once more began to decline.

Well, Elizabeth was deeply curious about this. So she decided to wait until after the hospital had closed down for the night and hide out and see if she couldn't figure out what was happening. It took a few days but she finally realized that one of the cleaning staff, an old, heavy-set black woman (who as it turned out had only an 8th grade education) was spending time in each of the patients' rooms. And it was *only* those patients who were getting better. Tellingly, the woman had weekends off.

Finally, one night, after Elizabeth was certain the woman was the cause, she decided to talk to her. So, as the woman left one of the rooms and began rolling her cart down the hall, Elizabeth came up behind her, touched her on the shoulder, and – when the woman turned around – said, "I'd like to know what you are doing with my patients."

The woman flinched, obviously afraid, and said, "Nothing! I'm just cleaning the rooms, ma'am. That's all I do. Now I got to get on with my work." And she began to move away again.

Elizabeth reached out and carefully touched her arm once more. "I'm not mad; you're not in any trouble. I just want to know what you are doing when you are in there. Whatever it is that you are doing, they seem to be getting better because of it."

The woman calmed a bit, then looked at Elizabeth more closely and said, "I just sit with them and hold their hands. That's all."

And here she paused, then after a moment continued, "It's hard to die alone, to die without love around you. I know for I have had my babies die in my arms, and my brother, and my parents, too. It's so very hard to die without love." And then she looked at Elizabeth even more closely.

"It is so very hard to die without love . . . but it's even harder to live without it. What I do is I just sit with them. I hold their hands and I listen to them. And as I do, I love them." She gave a sort of soft, sad laugh then, filled with meanings it took Elizabeth years to fully understand.

"They don't get much of that around here, you know," the woman said, glancing around the hall of the hospital. Then she smiled at Elizabeth and turned away and went on with her work.

Elizabeth said that much of her motivation for creating the hospice movement in america came from this moment. After the encounter, she spent a lot of time thinking about how the dying were being treated. She began to see the hospital and the medical system with different eyes. She began to see her patients with different eyes. And herself.

Then, slowly, haltingly, she started to put love into her relationships with them and their families. It was uncomfortable at first, and very awkward for a person born and raised german-swiss as she was. But she learned, over time, to just let herself love them . . . and importantly, to allow herself to receive their love in return. And as they shared this love together she paid careful attention to what happened as they did. The terminal patients she worked with still died but *how* they died and *how* they lived until they died was very different than it had been before.

As she said later in her book *To Live Until We Say Goodbye*, the one filled with photographs and stories of the dying, these photographs and stories are "a memorable symbolic language" that reveals "what it is like to go through the struggle, through the pain, through the loneliness, like a rough rock that is put in a tumbler" and emerge on the other side "with the radiance of a jewel."

> *So it is the purpose of this book to show what can and will happen to human beings, young and old, child and adult alike, when they are in the process of being destroyed by a malignant growth and yet can emerge as a butterfly emerges from a cocoon with a sense of peace and freedom, not only in themselves, but in those who are willing to share their final moments and who*

> have the courage to say good-bye, knowing that every good-bye also includes a hello. . . .
>
> It is fear and guilt that are the only enemies of man, and if we have the courage to face our own fears and guilts and unfinished business, we will emerge more self-respecting and self-loving and more courageous to face whatever windstorms come in our direction. As one of my teachers so beautifully put it, "Should you shield the canyons from the windstorms, you would never see the beauty of their carvings."
>
> It has been our life's work to help our patients view a terminal illness not as a destructive, negative force, but as one of the windstorms in life that will enhance their own inner growth and will help them emerge as beautiful as the canyons which have been battered for centuries.

Elizabeth worked hard to teach these truths to the hospital administrators and to the doctors and the nurses who worked in such surroundings. Because what is true is that those of us who work closely with death and dying and loss are also being battered by one of the great windstorms of life. It cuts into the foundational stone of the dying as well as their families – but it also cuts into ours, into the deepest regions of who we are. We too are shaped. And we can become more than we were by that shaping or we can merely become deformed.

The hospice movement came from Elizabeth's work and it has helped bring a different way of death and dying to the terminally ill and their families. But the hospitals and doctors and nurses that she so wished to help have benefitted little from her teachings. For the industrialization and corporatization of medicine and dying have kept her most essential teachings sequestered, far away from one of the places where it is the most desperately needed: the hospitals. There is, regrettably, not enough profit in love and compassion and so it has found no place in a rational, mechanicalized medicine. It does not enter into its approach to life . . . or to death.

Elizabeth knew firsthand, and insisted it was necessary, that those of us who, as a way of life, enter the territory of grief and dying must have the touch of love upon us every day. She insisted that doctors and nurses needed rooms where they could go to process loss, where they could come to terms with the emotional impact of the dying. Places they could be touched and loved and weep when it was necessary. For she knew the most important of all lessons: that those of us who work with dying and severe loss walk a road filled with feelings, terrible grief and fear among them. Without love, she said, none of us can survive what we find there. And as she told us once, as we were weeping from the impact of her teachings, "Now you are *feeling* like human beings, not acting like dispassionate scientists."

She told us, though it took time for me to believe her, that after our time with her, we would need to create or find our own places where we felt loved and nurtured, where we could recover from what we go through as we travel this road. Places where we are held and listened to, places where we matter and know we are not alone. Places filled with aesthetic dimensions, like the temple that Joan Halifax built at Upaya. Places in wilderness, too, that touch the depths of us. Places where love resides. Elizabeth wasn't sure we could remain sane, or whole, or healthy without having those things.

I have learned after a half century on this journey that she was right.

I have told this story of the janitor many times throughout the years of my teaching. And what is odd about it is that most often when people mention it to me, they say, "I loved that story you told about the nurse sitting with the dying people in that hospital."

And as I always respond, "Well, no, it wasn't a nurse and that's the point of the story. It was an old, heavy-set black woman who had no more than an 8th grade education. It wasn't a licensed professional who knew the importance of love. It wasn't someone who used big words and believed that emotional distance made their work more legitimate. It wasn't someone who had been trained out of their capacity for empathy, their capacity for compassion, their capacity to feel. It was just a person who had chosen to retain what made her most human and to then touch those who were lost and abandoned and dying with her humanness."

The attainment of an advanced degree – as I have found to my dismay – is not often survived by the person who began the schooling to get it. Someone with the same name continues on . . . but they are not at the end the human being they were when they began. Quite often the most important parts of them, and of their humanness, don't survive the journey.

It is in the chambers of the uncorrupted human heart that truths like this live – and from which they emerge. It is not something that any school cares to teach, or any hospital cares to make a cornerstone of their work. It's something that has been abandoned, considered unimportant, no matter how much research continues to show that it *is* important and essential to the healing of what plagues us.

None of the work for which Elizabeth is known came from the halls of the academy. It came from doing exactly what she was told she should not do. It came from her visits to the concentration camps after World War II, the places where she experienced the crimes that rationality had committed in the name of science and health.

It came from the young woman she met there, Golda, whose family had died in the gas chambers at Majdanek while she, for no reason she could ever fathom, survived to see their ashes float up later that day from the smokestacks. For it was Golda who replied, when Elizabeth asked, "How can you be so calm, so free from bitterness, after what you experienced here?" (and changed the world when she did), "Because the Nazis taught me this. There is a Hitler inside each of us and if we do not heal the Hitler inside ourselves, then the violence, it will never stop."

And Elizabeth's work came, too, from Golda's answer to Elizabeth's next question, "What are you doing now?"

"I am working in Germany," Golda said, "at a hospital for German children injured during the war, the children of the Nazis who sent my family to Majdanek."

Elizabeth, shocked, asked her why she would do such a thing. "How else," Golda replied, "can I heal the Hitler inside me but to give to them what they took from us?"

Elizabeth's work came from the truths with which Golda touched her during an incredibly difficult moment of personal pain as she stood in the ruins of that concentration camp and found her internal assumptive world crumbling within her. It came from listening to the dying and their families, from an old black woman with an 8th grade education. It came from her entering the territory of death and dying and finding out what was true for herself. It came from her learning to grieve, openly and honestly. It came from her willingness to love . . . and to be loved in return. It matured as she, over time, learned to be unafraid of death, of suffering, of the dying of those she loved and cared for. And it flowered as she found her essential self and her life of service in the one place she thought it least likely to be found: in the midst of loss and death and endings.

She did not find despair or apathy to be the final state of being that occurs when a person truly accepts the fact of a terminal illness. She found it to be something else entirely.

CHAPTER SIX

The Life We Live with Afterward

An overarching aspect of the trauma of genocide was that the violence was initiated by political and governmental entities, and carried out by individuals, whom the victims had reason to believe would, at the very least, act with a basic concern for human life. Beyond the terrors of the actual experience of violence, the victimized individual must also contend with the loss of faith in a protective world.

SAMUEL GERSON

[As] authorities over us are removed, as we wobble out on our own, the question of whether to be or not to be arises with real relevance for the first time, since the burden of being is felt most fully by the self-determining self. . . . I know of nothing more difficult than knowing who you are.

WILLIAM GASS

The Way It Is there's a thread you follow. It goes among things that change. But it doesn't change. People wonder about what you are pursuing. You have to explain about the thread. But it is hard for others to see. While you hold it you can't get lost. Tragedies happen; people get hurt or die; and you suffer and get old. Nothing you do can stop time's unfolding. You don't ever let go of the thread.

WILLIAM STAFFORD

This is not a book that talks of exterior solutions to the problems that face us. It is not a book about gardening or sustainable sources of fuel or what legislators need to do *now* so that the climate crisis can be avoided. It is not a book of techno-utopian solutions that will allow our species to continue on as it has been doing for far too long. This book is concerned with other things entirely.

Yes, as the title says, it's about the journey into and through grief and loss but it is also, importantly, about reclaiming what is most essentially human in us during the process. Specifically, reclaiming our capacities to feel the touch of the world upon us, to love, to care, to grieve, to deepen in wisdom and maturity and to make those things primary rather than secondary to a lived life. This entails a permanent refusal to perceive or define the natural world and ourselves through the lens of dissociated mentation and its various rationalisms such as economics, capitalism, and a reductive, mechanicalistic science. It is about accepting some rather harsh facts and about a journey that takes us from one way of being to another, one that necessitates the loss of the old internal assumptive world that we have carried within us for so very long.

Giving up that old, interior structure allows a very different orientation to be constructed, one far more congruent with the ecological realities of Earth – and our own, crucially important human qualities. And that is a necessity. For while it *is* true is that we need to change human behavior in the outer world, it is even more true that we need to change human behavior in the interior world. And that means shifting our internal paradigm and structures – our internal climate of mind. We need a sustainable way of life *in here* as well as out there. A way of being that is regenerative, that is oriented toward life, that reclaims what we have lost, including important elements of our own humanness.

Under the teachings of reductionists, corporatists, and monotheists, we have incorporated the same kinds of beliefs about ourselves that we have about the external world. And the consequences have been the same for us as it has been for the exterior world. As Jesse Wolf Hardin once put it, "I, too, like this land, have lost parts of myself." The problem is that so very few of us are aware of it . . . at least not until our lives begin to fall apart and we are forced to face our interior selves. It is then we find that there is also a need for an ecological reclamation of our selves. In truth, we need to adapt as a species in far more complex ways than most people realize.

Adaptation does not only mean learning how to make snow caves in the arctic to survive a winter storm. It does not only mean making better spears to fight off animal predators. Nor does it mean creating more technologically superior spears and snow caves. (*Inconceivable!* as that is.) It does not mean what most people think it means. And this includes nearly every scientist, environmentalist, and techno-utopianist there is. Far too many of them have been trained to focus solely on the outward, to believe in and act on the control and manipulation of matter, including the ecological systems of this planet. But human control of the natural world is and never has been possible. That is an ecological truth, a behavioral limit, an axiom that we transgress at our peril. As Vaclav Havel once put it, "There are powers here against which it is wise to not blaspheme." Acting under the belief that we can control nature is, in an ecological sense, in every sense actually, extremely unwise. The natural world is simply far too complex, ancient, and interrelated for rationality and limited human perspectives to fully understand.

This is why all attempts at control result in unintended and unexpected side effects. The more complex and far reaching the technology that is used, the more complex and far reaching the side effects. And those side effects will always need other interventions that then create more unforseen side effects that then need more interventions that . . . and it just keeps on and on and on and that is why everything always turns out so wrong. Natural systems are meant to run themselves through complex feedback processes that have taken billions of years to develop. Every time humans interfere, by building large dams for instance . . .

> *Which seriously biodegrade after sixty years or so, necessitating their removal and the dumping of the old concrete and steel, then rebuilding them to last for only another sixty years once more.*

. . . we interrupt those processes, interfere with critical feedback loops and their ecological functions. And things – always unforseen – begin to go wrong. We are neither smart nor wise enough to run the natural world. We can't even run the human world without seriously screwing it up. History is a very good teacher . . . if we are willing to pay attention to it.

Adaptation means *adaptation*. It means conforming ourselves and our cultures to the reality of those immensely large, ancient, and powerful ecological feedback loops and processes. (Something our hubristic belief in control has made impossible the past two centuries.) Now that Earth systems have begun to destabilize we are being *forced* to adapt, whether we wish to or not. Ramping up technological interventions in an attempt to control the destabilization of natural systems will only stimulate faster and more severe destabilization. What is necessary instead is to *adapt*. (Adaptation and control are not the same thing, they are antithetical to each other.) And that demands an alteration, a shift, in how the world outside us is viewed – and in how *we* ourselves are viewed. (And no, we are not a cancer or a virus.) It necessitates a recognition that the altered environmental states around us necessitate a shift in how we fit into that altered world. And this doesn't mean merely altering outward behavior . . . which is what nearly every climate writer is proposing we do. It means a significant alteration in how we perceive, experience, and are oriented toward ourselves and the ecological scenario from which we have come and in which we are inextricably embedded. That is, our entire internal assumptive world must change. It necessitates a *shift* in our ground of being.

There will be people, of course, who will read this book and believe that I am saying that no outward action can be taken to address the ecological crisis of our times unless months or years of psychotherapy (or some such thing) occurs first. (This is often one of the simplistic, default responses of those who are focused on the outward.) But I am saying no such thing. What I *am* saying is that to more properly address the problems that face us it is necessary that a deep and unremitting conversation between the self and the natural world take place – *continually*. We have to relearn how to fit ourselves into the natural world, to reinhabit our interbeing with it and its ecological processes. And this demands that we actually begin to listen to what we are being told. We have to give up our monologue, our telling the natural world the way it is going to be. We have to give up our human-splaining. And listen.

When a person actually listens to outside feedback – as any recovering addict will tell you – they begin to hear truths they were previously unaware of or did not want to hear. This, often for the first time, offers the possibility of real change. And change is a *process*, it takes time. We learn, we change, and we act while we are doing both. Over time, the work we do, the life we live, is refined just as our internal assumptive world is itself being refined and restructured. It is a one-step-at-a-time sort of thing. And we always have to start with where we are; there is no other place to begin.

In this instance, who we are listening to is not our family or friends but Earth itself. The adaptation I am talking about must of necessity be oriented around an interactive engagement, a dialogue, a conversation with Earth and its complex ecological truths. And this means recognizing that something's out there besides the human. A something that is far older and larger and more powerful than we are.

The truth is that we can no longer continue to think of ourselves as isolated consciousnesses on an inanimate, insentient, meaningless ball of resources hurtling around the sun. (Or nearly as bad: that we inhabit a kind of park that is here for our unlimited amusement.) That is not what this scenario in which we are embedded is – thinking so is the great, terrible error of reductionist science (which itself grew from the seed of a body/mind/spirit dissociation embedded within monotheistic beliefs).

This scenario is in fact a living, breathing, interactive field of meaning possessing extremely complex, inextricably interwoven-with-each-other physical and non-physical attributes. And while it really irritates rationalists, scientific reductionists, and monotheists to say this, among these attributes are consciousness, intent, and the capacity to communicate. We are only one particular expression of a much larger phenomenon. There is no attribute we possess that the scenario from which we come does not also possess. And because of this there is far more available to us here than rationalist dissociation or monotheistic either/or perspectives and dualities believe.

Thus, when we enter a meadow we also enter a climate of mind. For those of us who feel the touch of the world upon us – and are responsive to it – we take on the property of mind of the meadow as we walk and in so doing we perceive the world around us from a very different perspective. We *feel* differently, we are *oriented* differently, and in consequence we *perceive* differently. A calm pond lends us serenity. When its waters are disturbed by wind we also are disturbed. Our feelings do not come solely from inside us.

As James Hillman once put it, "Depth of soul lies not just in us; it resides in the planet's own nature." Restoring the bond between the two, between the inside and the outside, is no irrelevant matter. In many ways this book is about that necessity. *Restoring the bond between the human and Earth is in fact an essential ecological reclamation of the self.* And that ecological reclamation involves a decolonization of the soul, an un-domestication of personal identity, a freeing of mind from cliched, unexamined patterns of thinking. Integral to this is the restoration of the aesthetic sense. For only with an active aesthetic sense can we recognize and respond to the touches of the world upon us, upon our hearts, touches that are filled with complex meanings and communications that we, by our nature, are meant to receive and respond to. Restoring our capacity to feel (and thus reclaiming our aesthetic sense) is the way out of our dissociative dilemma. It is the first great act of disobedience.

Our ecological world is shattering, old forms are collapsing. From those scattered pieces Earth will create a new form – as it has so many times before. Is it really all that surprising then that we might be experiencing inside ourselves what Earth itself is experiencing? That we, who come from Earth, might feel in the deepest regions of ourselves the same shattering that the ecosystems and our kindred species are experiencing? Can it really be all that surprising that if Earth's climate is changing, its climate of mind might also be changing? That we might feel that alteration, take on that changing state of mind, just as strongly as we feel (and take on) the alterations in mood that attend wind alterations on the surface of a still pond? That as Earth's climate changes we find that our species' climate of mind is being forced to change as well? That all of us, along with Earth itself, are also, whether we wish it or not, inextricably involved in a process of adaptation and change?

The climate of mind inside us must of necessity adapt to the reality that surrounds us now if we are to learn again how to live sustainably on this planet we love. We no longer have a choice. We are out of time.

The ecological disruptions that are occurring are themselves putting tremendous pressure on our species' internal assumptive world. And as those internal structures collapse, social upheaval is inevitable. It is going to be a very messy and painful process. People, either as groups or as individuals, rarely give up their internal assumptive world until outside forces insist they do so. Even then they resist for as long as they possibly can.

What so many of us are experiencing now is the *actual* reality of the world, the harsh facts of life breaking through our habituated not-knowing, our domestication, nibbling away at the edges of our internal assumptive world and destabilizing it. It is the source of our fear, our unrest, and deeply connected to our grief.

This book is about those feelings and that experience, what happens inside us in response, and the journey into and through all that in order to find a different way of being. It is about *true* adaptation. From that adaptation we can become what some people call a "live third" or "an engaged witness." This is what Golda and Cousteau, Elizabeth and Stephanie Simonton, Vaclav Havel and Nelson Mandela, James Hillman and Robert Bly, Viktor Frankl and Masanobu Fukuoka and so many others have themselves become.

It is this particular state of being that I have been talking about all along. It is in this state that we are most able to live in "engagement with the fate of the unbearable," as the psychoanalyst Samuel Gerson once put it. Over time, the losses to which we have turned our face, the internal work we have done, including the destruction of our internal assumptive world and its rebuilding on newer, more ecologically (and human) accurate foundations, have allowed *grief* to penetrate every part of the self. We have given up the ascent of optimism and responded to the call to descend, to intentionally travel the road of ashes and loss. And during that traveling, oddly enough, grief and loss become leavened by crucially important attributes of self and character as well as the eternal truths of life that have been found on the way.

The state of being that emerges, and that this book is about finding, can *only* arise from deep loss. It is oriented around what some call *gravitas*, a word that has close relationship to, and the same linguistic root as, *gravity* and *grave*. Only through the encounter with severe loss and death can gravitas emerge. And with *gravitas*, always, **comes** wisdom.

> *grief is the soil from which wisdom grows*
> *it is the only soil in which it can grow*

With the emergence of gravitas it is no longer necessary to turn the eye or heart away from what is happening in the world. It becomes possible then *without losing the self or being overcome by the pain* to hold the immense suffering that is felt and seen all around us.

As suffering touches us, it simply deepens the *gravitas* inside, merely strengthens the keel that keeps the ship of life afloat. In fact, suffering transfers itself, and all the energy that accompanies it, into that keel. And the more intense the grief and loss, the more the keel is strengthened. So, no matter what the storms of life bring us, we remain afloat in their midst.

With this alteration of self it becomes possible to more fully take on the task that is in us to do in the times in which we find ourselves, the one that is integral to us, core to our nature, the one that one does "because one must." It is not something chosen but something that by our nature has chosen us. And it is all of us, in our millions, doing the unique and innovative work that is in us to do that our species needs now. Change always begins at the individual level and works its way upward. It is never a top down process. It is only when politicians see a huge crowd of people headed in a new direction that they are finally able to jump in front and say, "This way!" They are in actuality not very bright or honorable people. It is very rare for any of them to truly lead.

The task that exists for each of us, and that this journey fully brings forth, is rooted in our own uniqueness, the sense of the right that we have, and the urgings of our heart to take this path and *this* action rather than *that*. And, of course, it is also rooted in trust (including a trust of one's self). But most of all it's a trust in Earth itself.

Earth has been through worse than this many times before. And

Earth, despite the patronization of rationalists, makes no mistakes. Every species it generates out of the great ocean of being in which it resides is brought forth for specific ecological reasons – whether rationalists understand those reasons or not. It is foolish to think that an organism with a life span of eighty years can understand ecological movements that might need five thousand years to come to fruition. Or five million. Or five hundred million. Earth thinks on *very* long time lines. Rationalists and, regrettably, most environmentalists, do not.

We who have come to understand this and have made our peace with the time in which we live trust Earth. Our faith is rooted in Earth itself. In the life and renewal that come and have always come from endings and dyings. It is a faith in the way that every spring, plants emerge from the soil, in the way our hearts rise up and sing when we encounter the beautiful love and life in a bounding puppy, in the wisdom of stones and water and great mountains. It is faith, too, in the human species, in the children who will be born, in the millions of us who are responding to the touch of Earth upon us, telling us that there is something that we must do. It is faith in the movements of Earth, the ecosystems that will arise, as they have always risen over geological time after this kind of change has come. It is faith in the work we are called to do. And that work, and our service, is, and must be, oriented toward the long view.

For what we face now, whether we want it to be or no, is not a safe, predictable, controllable world. Quite the opposite. The human species, *as a whole*, has entered uncharted territory and it is a territory that we will be *in* for a very long time to come. Nevertheless, others have traveled this sort of road before us and their teachings are invaluable for what we now face. We are not the only ones who, on a large scale, have faced the necessity to reorient our internal assumptive world.

Samuel Gerson, when he talks about "engagement with the fate of the unbearable," is speaking about survivors of the holocaust. And what they experienced has great relevance to our time and what we now face. If you can read (and hear) the following words as if it is Earth and all the living beings of Earth speaking them, then you will catch a glimpse of what is necessary for us to become. As Gerson says, *it is in those who can do this* that life gestates and "into whom futures are born."

> *What then can exist between the scream and the silence? We hope first that there is an engaged witness – an other that stands beside the event and the self and who cares to listen; an other who is able to contain that which is heard and is capable of imagining the unbearable; an other who is in a position to confirm both our external and our psychic realities and, thereby, to help us integrate and live within all realms of our experience. This is the presence that lives in the gap, absorbs absence, and transforms our relation to loss. It is the active and attuned affective responsiveness of the witnessing other that constitutes a 'live third' – the presence that exists between the experience and its meaning, between the real and the symbolic, and through whom life gestates and into whom futures are born.*

We are again living through a holocaust but this time a holocaust of the world we have known all our lives. It is a holocaust that is killing by the billions the kindred lives that surround us. Underneath the behaviors that lead to that killing is a belief that those life forms are *other*, less important than the human, insentient, merely extractive resources or a necessary collateral damage to our "progress." It's the same kind of thinking (something Val Plumwood called sado-dispassionate rationalism) that led in its time to the holocaust of the jewish people. It is what has led to this holocaust that some are calling the anthropocene. Those who lived through that other holocaust and survived it – *and not all those who lived survived* – were transmuted by that experience, became under the pressure of what they endured, a "live third." They became, as people like Samuel Gerson have, an engaged witness.

It isn't easy to do. But it is what is needed now. Earth is asking each of us to make that same journey, to become an engaged witness to the living reality of what is happening. To become a "live third." And from that place, to respond.

I can promise you this . . . it won't be a boring life.

EPILOGUE

"It's Hollow Inside, Isn't It, Just Like Me?"

My elders have told me that the trees are the teachers of the law. As I grow less ignorant I begin to understand what they mean.

BROOKE MEDICINE EAGLE

Those we respect as our great teachers, from a certain distance, were faithful. They did not break faith with their beliefs, they remained dedicated to something outside the self. As far as we know they never became the enemies of their souls or their memories.

BARRY LOPEZ

I would do for you what spring does for cherry trees.

PABLO NERUDA

A young woman came to me long ago. She had heard that I knew something of plants and their medicine and because she was suffering she hoped I might be able to help.

It had seemed an average day when I awakened that morning. But it wasn't.

The girl hovered in the doorway, making it a central pivot around which her life's choices circled. Half in, half out, uncertain about committing. Hesitant when she entered.

I invited her to sit. She chose the blue chair. She perched tensely on its edge as if ready to flee. Her breathing was short and rapid, high in her chest. I offered tea, she said no.

She was young, perhaps twenty-eight, the lingering baby fat of childhood completely gone, the mature woman's bones in her face struggling to find expression. Her cheeks were drawn, the hollows around her eyes shadowed with pain. The eyes themselves, except when they, with jerky movements, took in my face and the room, were fixed and staring, the pupils contracted. Her clothes were plain: jeans, a workmanlike white shirt, sleeves rolled to the elbow, and worn, well-used running shoes. Her hair was a lifeless brown.

There is sometimes a porcelain whiteness to the skin of caucasians, delicate and soft. Life shines through that skin, luminous, like a candle flame glimmering through the tender shade of a lantern. But this young woman's skin was pale, as if never touched by sun, as if lived in darkness and such isolation that all life force had been drained away. It seemed artificial, an untended garment, unalive, unloved, untouched.

She was in-between. Between the lingering youth of her late twenties and the mature woman that comes into being, perhaps at thirty-five. She was caught between the end of one life and the beginning of the next. And as it is with all life transitions, in that liminal region between one stage of life and another, contractions had begun. They were compressing her, bearing down, demanding she move. But the intangible cervix within her, that is within all of us, had not yet dilated enough to permit birth. She was lost and afraid.

In a barely perceptible monotone she told me her periods were irregular with intense cramping and heavy bleeding. She said her hands were always cold.

Those hands lay on top of her thighs, near the knee, palms down, fingers slightly curled. They were pallid and slightly waxy like the skin of her face, their character only beginning to form. Like many young hands they were only partially alive, rarely inhabited. Her fingers, those sensitive touching extensions of herself, possessed nails and cuticles that were chewed and torn. As if she were turning inward and tearing away at herself. Eating herself alive.

The muscles of her forearms were taut, filled with the tension of barely holding on. And, occasionally, as we talked, her hands would tighten spasmodically and, finding nothing to grab hold of or no reason to become a fist, they would relax again. But . . . half open, curled, ready.

I asked her to tell me more about herself and she said she was in the midst of a messy divorce, that she couldn't sleep. And while she talked I noticed she sat slightly hunched over so that her chest was hollowed. As if she were trying to make her breasts smaller. As if she could call them back inside her, make them disappear. As if maybe her shoulders could curl all the way around her then, cover her heart with muscle and bone and strength of arms. As if her heart were weeping and she were trying to comfort it, trying to protect it from the world, from ever being hurt again.

I was beginning to breathe shallowly too, my body tensing up, eyes becoming pinpoint focused when suddenly into my mind came an image: *Angelica*. And I took a breath and let it all go.

I told her there was a plant I thought she should meet and so we went for a walk.

I took my staff but the sun was warm and the sky clear – I did not need a coat. We cut across the meadow in front of the house, skirting the driveway with its hard-packed gravel, and began wandering down the other side of the ridge. After a while the slope dropped rapidly and I angled left, taking us into a gentler-sloped ravine, one shaped over time by water running off the ridge. It had silted up over the centuries, becoming a grassy, slightly-sloping flood pan with occasional wild rose and spreading, prickly-thickets of the short, matted *ceanothus* known as red root. After thirty yards or so the ravine spread outward, becoming a broad open meadow, an aspen grove at its base.

The meadow's lower edge was slightly marshy and here the milkweed grew, butterflies thick among them. Scores of the plants made their home there, taking advantage of water seeping upward from the underground aquifer, itself kept rich by snowmelt from the high ridges of the mountain behind us. We moved carefully through the milkweed, then stepped into that gentle, luminous world that only aspens and their groves create.

The air of the grove was thick with aspen's peculiar resinous smell and the hum of bees busy harvesting its sap to line and protect their hives. The small, heart-shaped, slightly serrated leaves, even on trees deep within the grove, were moving in the almost imperceptible breeze. They were lightly shaking, almost shivering, in unison as if they were dancing with some secret joy known only to themselves. I could feel my spirits lift, my feet dance/walking to their secret joy as I moved among them. And I saw the young woman breathe a bit deeper, too, saw her begin to relax, the gladness of aspens finding its way even into her difficult interior world.

Aspen roots are some of the oldest living beings on Earth, over 100,000 years some of them, but the trees they give birth to remain relatively young. They rarely grow more than a foot to a foot and a half thick before dying back. This grove was a small one, only about forty feet in diameter, and as we moved down the hill the bright dancing light of the aspens soon gave way to the deeper and darker fir forest that lay below.

The firs here were much older and larger, perhaps two or three hundred years of age. Their numbers increased as we walked, the forest thickening around us until the sun's dappled golds only occasionally slanted down through some opening in the overstory. It was darker here, more peaceful, silent, slower. Almost mythic in its feeling – as ancient, elder forests often are. Our breathing deepened further and our minds slowed, then calmed, as we took in the trees' sun-quickened, musky scents, as the silence and age of the place penetrated us, as we took on its climate of mind.

Soon, we neared the small, year-round stream that wove itself through the firs, rocks, and soil at the bottom of the sloping mountainside. Cow parsnip began to appear, big elephant-eared leaves spreading out from strong muscular stalks. Then . . . a profusion of plants: violet and figwort, dogbane and osha, with horsetail and poleo mint along the shallows, scattered amidst the rocky sand of the stream.

The north-facing slope on the other side of the stream was plant-thick, old, and deeply green. The firs there were tufted with short, gray-green growths of the lichen some call usnea. A few were completely enveloped, the trees become shaggy, shambling green beings, thirty to fifty feet tall.

I could feel them stir, wakened from their slow dreaming by the touch of our seeing, the feelings of our hearts for them, by our presence in their midst. Soon, we came upon a great rocky outcropping concealed within the gloaming mystery of those ancient trees. It towered over us, thirty or forty feet high, sides covered with lichen and shadows. At its base grew a tangle of wild roses and mountain mahogany.

There was an animal trail at our feet, just this side of the stream. We turned left, moving with it as it followed the twisting and turning of the water. Occasionally we would happen on huge rotted stumps torn open for grubs by the bear that lived on top of the mountain. Scattered clusters of false solomon seal covered the ground, dug from the base of the stumps where bear had left them. They were not yet wilted, the roots long, pale, knobby, almost translucent.

The trail was only six inches or so wide, its soil black and rich, the plants along its length emerald green in that shadowy velvet world, We sank down slightly as we walked, the soil springy, alive, moving in time with us, like a trampoline almost. The two of us were held, suspended over its ancient depths, like water spiders skating along the surface tension of the soil. And then, ahead of us, I caught sight of the unique angelica that grows only here in the high rocky mountains. It's unlike any other species of angelica I have known.

Angelica was growing on the other side of the stream, snuggled up against the north-facing slope and those ancient firs. It was right at the water's edge, three or four feet off the trail from where we stood. Majestic and stately, vitally alive and deeply feminine, six feet tall, perfectly balanced between Heaven and Earth. I watched closely now, to see if what I suspected would in fact occur.

The young woman was a step or two ahead of me and slightly to my right. Ducking under a leaning lodgepole pine, she caught a glimpse of the plant from the corner of her eye. She stopped suddenly, as if some invisible hand had laid itself upon her, then she turned toward it and paused.

Something inside shook itself, visibly wakened, took breath, and stepped into this world for the first time. She began moving then, almost as if in a dream, feet sleep-walking of their own accord, an invisible force pulling the two of them together. When she came close, her hands began to flutter over the plant's leaves, barely touching, as if they were the body of a lover. She turned her head toward me, eyes questioning.

"What is it?"

"*Angelica*," I said.

She turned back, hands still fluttering, lightly touching. Then suddenly, she paused, took in a shocked, sharpened breath, said, "It's hollow inside. ... Isn't it? ... Just like me."

She looked then, to see my face.

"Yes," I replied. "Yes, it is."

Our eyes met, hers expectant, almost-hoping.

"Ask her to come into the hollow place inside you," I said. Then there was a pause and I watched while she did that thing.

She closed her eyes, her breathing slowed, deepened, stopped. There was a moment of waiting, of simple being, then a long ... slow ... deep breath filled her chest. The curled hollow disappeared. She straightened, her shoulders came back, her breasts reached outward, into the world. She stood taller, more balanced, fully erect now. Her skin flushed rose ... and came alive. The tension in her forearms relaxed ... softened ... released. She opened her eyes; they were moist and soft-focused, gleaming gently amidst the gloaming of the world.

"Oh," she said. "Oh." Her voice filled with wonder.

"Now walk awhile," I told her. As she did, moving slowly along the stream, her body continued to strengthen, straighten, fill out, mature before my eyes. I watched and saw that in some strange, magical way this plant was teaching her how to become a whole, mature woman.

For the first time in my life, I knew with certainty that all of us are born with holes inside us. And those particular holes do not come from wounds given us by our families or our culture or other human beings. They come into the world with us. They are holes that can only be filled by some wild thing from Earth itself. And they each have a particular shape to them – of plant or stone, of tree or bear. They are an emptiness meant to be filled by the kindred beings who companion us on this Earth. And without our allowing them to take their rightful place within us we live a half-life, never becoming fully human, never becoming healed or whole or completely who we are. Never becoming completely sane.

And I wondered then, what will happen to us as a species when we can no longer find these wild places, when the kin with whom we share this planet are gone. What will we do when we alone remain, orphaned amidst the shattered world the extractors have left behind them?

> *Long ago I was told – but never really understood – something that the elders of this planet have long said we must remember. In ancient times, when people first came upward out of Earth they were ignorant, without language, without law. The kindred nations who'd emerged long before our species saw and took pity on us, taught us how to be human, taught us language, taught us the law of Earth and how to live in harmony with all life.*
>
> *Over the years, as I have become less ignorant, I have begun to understand what my elders were trying to teach me so very long ago.*

After a while, the young woman and I walked on. We found another angelica further down the stream. Then she laid her hands along the leaves of the plant, and saying a prayer for the taking of her life she asked for *Angelica*'s help in the healing of her body. Then the young woman dug the plant, for it is the root that is most often used for medicine.

The root is delicate and somewhat small for such a tall plant. There's a central bulbous taproot, in shape and color something like a parsnip but smaller, with a whorl of smaller rootlets spreading out to hold the plant stable and upright. The color is tannish, not as white on the exterior as a parsnip, the interior a cream with a sublime smell, indescribably *Angelica*, aphrodisiacal. Human catnip.

The young woman kept the root close to her, rubbing her fingers along its sides, periodically lifting it to her nose as we took the long walk back to the house. For a month afterward she took a tincture of the root (which tastes much as it smells). Her menstrual periods normalized, the painful cramping slowed and stopped. Her body changed. And she changed, too – in more ways than I know how to tell.

Rationalists, reductive scientists, and nearly all monotheists will, of course, not understand this story. They will think it foolish, reject it and its implications. They will insist that it is "irrational, inappropriate, anthropomorphic." Most will angrily assert that it is just a projection, an animist tendency that only little children and the uneducated or primitive among us still possess, a holdover from our ignorant past and ways of living, when people who didn't know anything about the real world just made things like this up, before they knew of science or real religion or the modern world. They will insist that such an event, in reality, cannot and does not exist.

But every person who has retained their aesthetic sense, their sense of childlike wonder, their *humanness*, and who still remembers that the world is alive and aware and caring, will immediately understand the story. They know there is an important truth to it, that such moments are essential to our humanness, that more goes on here than rationalists and monotheists insist.

There are hundreds of millions of people to whom the world speaks in this way. They are people who still retain the capacity to learn from the wildness of the world and whom Earth teaches every day of their lives. Many of us on this planet still live as people always have, immersed in the fabric of Earth, immersed in its metaphors... and its mythic dimensions.

I know, for I have taught thousands of people in countries all over the world. Over the years hundreds have come up to me afterward and in small, hesitant, uncertain voices said, "You know, I have felt that way, too, as if the world had reached out and touched me. I have had experiences like that young woman and I thought for most of my life that I was crazy. Thank you for telling these stories out loud."

Colonization has been going on for a very long time, you know. It's just that some of us have been colonized a great deal longer than others. So much so that we no longer remember that we have been colonized. Nor do we remember what and who we were Once Upon A Time. Nevertheless, we are not, and never have been, isolated intelligences on a ball of resources orbiting a star. We are of Earth. We *are* Earth. And we have far more in common with a plant than a car. The mechanomorphism of the rationalists and reductionists is a projection . . . and it always has been.

As I near the end of the story I am telling you in this book, so many thoughts occur to me. They arise of their own accord out of memories I have long held within me, seemingly irrelevant perhaps, but then again, perhaps not. . . .

For instance . . . japanese physicians often prescribe *shinrin-yoku* (forest bathing) to their patients instead of antidepressants. Blood pressure lowers, heart function is enhanced, skin tension decreases, eye focus softens, breathing deepens, depression quite often disappears.

Funny that. Funny that, as James Hillman once curmudgeonly responded to psychotherapists who were arguing about what the most supportive office furniture and colors were for their patients: "The healing environment is the *actual* environment. Why don't you take your patients for a walk? Why don't you take yourselves for a walk?"

When we walk in wild landscapes we breathe in a complex mixture of compounds. Reductionists, as they always do, focus on the most obvious, the most simple of them. They continually talk about the oxygen/carbon dioxide exchange. They insist the other gases in our atmosphere are irrelevant. But of course they are not.

Plants are the most adept chemists on Earth. As David Hoffmann once said, "They have the ability to produce an almost endless number of chemical variations on a single chemical structure." They create both old and totally-new-to-this-planet chemical compounds throughout their lives, every day in fact. And they are far better at it, and far smarter about it, than we are. They are Earth's living, highly adaptable and intelligent chemists, endlessly recombining Earth's molecular structures in response to trillions upon trillions of subtle ecological signals. It is an integral part of their ecological function to do so. And they have been doing this for a very, very long time. Among the millions/billions/trillions of compounds they create are the highly variable and complex grouping of aromatic hyrdocarbons they continually release into the air.

When we breathe them in, the complex microbiome in our lungs metabolizes them, just as the microbiome in our gastrointestinal tract metabolizes molecular compounds from our food. In other words the microbiome in our lungs eats volatile, aromatic chemical compounds. Those plant-created hydrocarbons are essential to our health, to the microbiome of our lungs, which, like the one in our gastrointestinal tract, has pervasive impacts on our neurological and physical functioning. Playing havoc with current reductionist beliefs, it is when we inhale these aromatics in their incredibly tiny parts per billion or trillion that they alter our physiology and functioning the most.

Regrettably, the inhabitants of our lung microbiome do not make distinctions between the kinds of hydrocarbons they will ingest. They can and will metabolize any aromatic hydrocarbons that we breathe in, irrespective of what they are or where they come from. The chemically-modified, human-created hydrocarbons that humans generate from ancient plant-generated petroleum extracted from deep-Earth will do just fine. (Such as those from gasoline and cleaning products, perfumes and oils, paints and finishes, from a thousand other things we encounter every day of our life.) Regrettably, these artificial compounds aren't very healthy for us or our lung microbiome. They should more properly be thought of as aromatic junk food. And while it is becoming apparent to most people in the west that junk food alters the gastrointestinal microbiome in ways that cause disease, including deleterious functioning in our brains, it is not yet commonly recognized that the same thing happens with the microbiome in our lungs.

There are so many different dimensions, so many aspects to the ecological breakdown of our world. Hundreds of them, thousands of them. The vast majority are unknown to most people, including most scientists. Discipline boundaries have a great many severe side effects, a critical one being the high levels of ignorance amongst the overly schooled. Few of the extensively-degreed have the capacity or the willingness to look across multiple discipline boundaries to find out what is on the other side of those fences. Their perspective is always constrained by their training and focus no matter how many advanced degrees they have.

Still, there's another problem that they and all of us are faced with every day of our lives. It's pervasive in the ecological information that floods the media and scientific journals. And it is this: it's always one-sided. *It's always about us.* No one really looks at the other side of the exchange. For instance, while we walk in wild landscapes and breathe in complex plant hydrocarbons, the *plants around us are breathing in more than our carbon dioxide. They are breathing in the volatiles we exhale.*

Plant leaves have innumerable mouth/lungs on the underside of their leaves called stomata. And inside the stomata there is a microbiome too. Plants breathe in our carbon dioxide and breathe out the oxygen that we, and most animals, need to live. Their exhale is our inhale, our exhale their inhale. The great breathing of the world, of Gaia, of Earth. But inside our exhalations there is far more than carbon dioxide, there is a complex mixture of volatile hydrocarbons given off from the organisms in our lung microbiome. There are other compounds as well, transferred from our organs, through our blood, then into and then out of our lungs. Plants breathe in not only other plant-generated aromatic compounds but also exhaled compounds from a large variety of non-plant species, including us. Their microbiomes metabolize those compounds just as we do theirs.

> *Even more disturbing to reductionists, when plants sense that a member of their community is ill – including us (for they analyze the aromatics we are breathing out and from that can determine our state of health) – they alter their molecular recombining to generate compounds that will help move our bodies toward health. One of the plants' main functions is to monitor and maintain ecological homeodynamis within the regions in which they live. And that includes the healthy functioning of every organism that lives there, including us.*

Of course, just like us, plants are also breathing in industrially-created hydrocarbons. No one has taken the time to explore the alterations that occur in plant microbiomes as a result. Researchers have barely examined what happens to ours – until fifteen years ago or so, despite rather common evidence to the contrary, medical researchers and physicians insisted that human lungs were sterile environments except during infections. (Inaccurate knowledge among the overly-schooled is, contrary to popular assumptions, common and immense.)

Plant microbiomes, just like ours, do not make a distinction between manufactured hydrocarbons and those natural to life. And I often wonder what happens to them as they breathe those things in. I wonder, too, what happens to plants who continually breathe in carbon monoxide from cars. I know what happens to our bodies when we do; I wonder if it does the same things to the plants that surround us and upon whom our very existence depends.

No one that I know of has looked at any of this, has asked themselves, "What happens to the plants when they breathe in the crap that industrialists and corporations have created?"

All this mental analysis. All this chemical this and that. How foolish it is. All we come round to at the end is where we began, a truth known to all humans in all times that have ever been (except for ours, I guess): that a walk in the forest is good for us, for our health, for our hearts, for the soul of us. Why should we need reductive, mechanicalistic scientists to verify these things for us? How did they become the arbiters of truth? Of what is sensible? Of what our relationship to ourselves and the world around us should be? Of our humanness? Why is it that nothing that human beings know or have known throughout our time on this planet is valid unless some dissociated, rationalist person in a lab coat verifies it? How is it that have we lost ourselves so completely?

Those scientific stories about plants, their chemical abilities, the interrelationship between them and the other life on this planet? They are tremendously dissociative despite how intellectually interesting they sometimes are. As you listened to those stories? Did you begin to lose touch just a bit with what is most human in you? Did your feeling sense decline as your intellect became engaged? Did you notice that within all that chemical and physiological information caring and love tended to be absent?

There are *other* kinds of science, ones that are just as intellectually rigorous as that of the reductive mechanicalists, of the rationalists. They are ones that, *as you do them*, enhance and strengthen the human qualities I am speaking of in this book. As well, simply by engaging in them, those who do so are more firmly embedded within the ecological truths of this world. It is time, past time, for us to understand this and choose a different path, a different way.

Importantly, those reductive concepts, the way they are usually presented and generated? They come from a particular map, one common in our world, especially in the west. But it is only *a* map, not **the** map. And that reductive scientific map? It is only *partially* relational to the reality it points to. It is *not* reality, *not* the territory, *not* as comprehensive as reductionists make it out to be. Forgetting this is incredibly dangerous. It is part of what has gotten us into this mess to begin with.

Life is *out there*, in the living world, not in those mental concepts, not in that map, not in rationalist laboratories, not in that way of seeing the world. You can look all you want but you will never find love in the reductive rationalisms of science. You can look all you want but you will never find compassion in mathematics or ethics in physics or the wisdom of gravitas in botany. And if these things can't be found in those places, then there is something wrong with those places.

Out there, in the living world, is where we actually live, where all of us, including our kindred species, *actually* live. It is where life itself happens. It is far, far more complex, interesting, and mysterious than the reductionist map will ever understand. And I will tell you something now . . . *it is a place of feeling*. It is the place where love lives, where caring exists, where grief resides. It is where *life* has its home.

There are truths there that scientists will never find, truths that speak to what is most human in us, truths about this journey that all of us are on as we travel through life. I will tell you one of them now. It starts with a story that Henry David Thoreau told long ago.

Here is how it begins . . .

> *Everyone has heard the story which has gone the rounds of New England, of a strong and beautiful bug which came out of the day leaf of an old table of apple-tree wood, which had stood in a farmer's kitchen for sixty years, first in Connecticut, and*

> *afterwards in Massachusetts, – from an egg deposited in the living tree many years earlier still, as appeared by counting the annual layers beyond it; which was heard gnawing out for several weeks, hatched perchance by the heat of an urn.*

Thoreau goes on to say that there is a some essence inside us that emerges with us when we are born, an identity at our core that is uniquely our own. Far too often, as we move through the years, it becomes buried under the woodenness of social life and cultural expectations, under the teachings we receive as we grow, the maps we accept as reality and in which we try to constrain ourselves. As Thoreau then says, "Who knows what beautiful and winged life, whose egg has been buried for ages under many concentric layers of woodenness in the dead dry life of society" exists in each of us?

Later in life, perhaps some warmth (the metaphorical coffee pot – or urn as Thoreau names it) touches us and awakens us from our sleep. The deeply buried part of us begins gnawing at the woodenness that surrounds it, making a sound that family and neighbors can hear for miles. And one day, perhaps years into the process, a new self comes into being, one that looks and acts very differently than the one people knew before.

When we are touched by a living, caring intelligence from the world, a part of us, long buried and forgotten, awakens and comes alive . . . just as happened with that young woman. Her suffering, you might say, and the complaining it caused, was the sound of her beautiful winged life gnawing at the woodenness which constrained it. It kept pushing and gnawing and gnawing and pushing, and one day, when she least expected, it burst out of those constraints, taking her into a new kind of life. Importantly, the solution to her suffering was not something that could be found among the courtiers of power, in the academy, in physician offices, or in hospitals. It was at ninety degrees to her old way of being, to that wooden cultural world. It lay *outside* it. Out there, in the wildness of the world. In a place the domesticated and degreed rarely travel and, which, for sure, they don't understand.

Unsurprisingly, her suffering was coming in large part from the constraints, paradigms, beliefs, and structures of the human and cultural world in which she lived. A world that has, as Gregory Bateson put it, lost its sense of aesthetic unity – and its ability to sense and interact with the invisibles, the crucial intangibles with which we are surrounded every day of our lives.

There is a substance that is *in* the world that is possible for us to feel. It comes from Earth, from life itself. It is a soul essence, integral to life on this planet. Prior to the loss of the aesthetic sense, the majority of people on Earth, whether they talked openly about it or not – as they moved through the natural world hunting, fishing, farming, creating, living – knew of its existence, continually felt its touch, consciously included it in their work and lives. And for millennia, people used their aesthetic sense to shape and nurture that substance which resided deep within the wood and clays of the world – just as the those artisans did at the Upaya zen temple. You can *feel* it, alive and resonating, in their homes, their furniture, their art.

Thoreau and Emerson, Goethe and the ancient chinese poets, talked of that soul force. It wasn't an alien concept. It just was. A part of life itself. But as the aesthetic sense was lost, as it was abandoned for modernism, and later, when brutalism became the touchstone of our homes and lives, western humans became ever more embedded in the woodenness of rationalist and reductive materialisms, of literalness, and linearity. And the homes we lived in afterward? The furniture, the art? It became increasingly devoid of that essence, the soul of the world which our aesthetic sense is meant to perceive. Our homes became geometric boxes, cheap to make, terrible to live in, demeaning to the human spirit. By their nature, they urged us, at a level deeper than consciousness goes, to take on their climate of mind and thus to live a dissociated life

So it was that the young woman in my story became surrounded by a particular kind of emptiness, by a particular kind of lack, by a particular climate of mind. And she took it on as her own, as so many of us inevitably do. (As I myself did in my youth.) So much so that she did not even know what it was that had happened to her. She knew she had within her not only physical damage but a soul sickness as well. And despite her sleep, despite her pain, despite her wounds, when she was touched by the living being that *Angelica* is, by the soul and caring of that plant, something came in from somewhere, a substance and presence she had forgotten existed flowed into her. And that changed her, healed her in ways that reductionist medicine and science are unable to understand or even accept.

The healing environment is the actual environment.

Interestingly, in all the years I have told the story of the young woman no one has ever asked me what *Angelica* got from the exchange. (Things are, as usual, all about us.) But it is crucially important to understand that, as my beloved partner Julie McIntyre once put it, "It's their adventure, too." They *need* us, just as we need them. They get something from you and me and all of us just as we get something from them. There is an *exchange* of soul essence, an intimacy that occurs. Something from inside us crosses the distance and enters into them, just as something from them crosses that distance and enters us.

This sort of thing is always denigrated by rationalists, rationalists of every kind, shape, and form. Nevertheless, *all of us*, one time or another, have had the experience ourselves. Like when we see a young puppy across the room and, caught up in their puppiness, unable to restrain ourselves, we say, "Here boy, here boy." And pat our leg.

The puppy looks up and sees us and an expression comes over his face that says, "It's you. *It's you.* I have been looking all over for you!" Something comes out of him then, something we can *feel*. It crosses the distance and enters inside us. And at the same time something leaves our body, crosses the room, and enters the puppy. We know it the moment it happens. *And so does the puppy.* His whole body begins to wag, he bounds over and we can do nothing then but pick him up and set him on our lap. Our hands touch him then, seemingly of their own accord. Just as his tongue touches us, caressing our skin. A communion takes place. One essential to our humanness, to the soul of us.

The ancient athenians had a word for this experience of intangible, non-kinesthetic touching. They called it *aesthesis*. It is the moment when a soul exchange occurs between a human being and something from the wildness of the world. Always, at the moment of encounter, there is a slight pause as the impact of the other being is felt. Then a sudden, deep breathing-in takes place as the exchange occurs – an inspiration. (The athenians were very clear that inspiration comes from the *world* and not from inside the person.)

When we unexpectedly come upon a great rock formation or an elder tree, we *feel* the *presence* of something alive and strong and old. Something outside the human. And for a moment we are *held* in the feeling of that presence . . . but, at the same time, we are also *beheld*.

And very quickly, there is that intake of breath, that inspiration. We breathe them into us and they breathe us into them. This is not a river, it does not travel in only one direction. They need us just as we need them. We are all in this together. We are not alone here.

As we travel through these difficult times it is important to remember that. For the wild things of this world are working just as hard as we who care for this Earth are working. They are responding to the alterations in climate, imbalances in Earth functioning, just as we are. What is happening matters to them, too.

They are innovating, just as we are, crafting solutions, just as we are. And despite all the phytohysteria about it, the great movement of plant populations across the surface of this world is an integral part of their response. They are not merely invasives. Not insentient beings at the effect of forces they cannot understand. Not invaders destroying our lives and gardens. They are altering their forms and behaviors, moving in response to forces rationalists do not understand, entering into damaged ecosystems (no matter how green those places appear to the human eye), and responding to Earth's destabilization. In those new landscapes they make their thousands and millions and billions of chemicals – those alkaloid verbs, tannin nouns, mucilaginous adjectives – each a part of their complex conversation with the world. And this conversation is modulating the shape, form, and behavior of damaged ecosystems – and every organism that lives within them – moving them, and those ecosystems, back toward health – toward a more functional adaptation in the face of what is happening.. Every action they take is specific to the ecosystems they enter and the damage they find there. (And all those bacteria? They are working as hard as they can as well.) We are *not* isolated intelligences inhabiting a ball of resources hurtling around the sun. We are companioned by innumerable kin who are just as concerned as we are about what is happening. And those kin? They are very, very smart. They have been around a long time; they have seen much worse than this. We are not doing this work alone. The healing does not lie upon your or my or any person's shoulders alone.

Maybe knowing that will help you in the times to come. It is part of that faith I was talking about, the faith that is hope. The faith that Deborah Bird Rose talks about in this way:

> *[This kind of] faith is not defined solely by cognition; it can be located throughout the body, and it may often erupt mysteriously, being called into existence by what is outside us or precedes us. Faith, in my view, is action toward intersubjectivity. It is called forth by that which is beyond the self and thus equally is action arising from intersubjectivity. . . . I know that my life takes a twist into life-affirming action when I ground my life's work in intersubjectivity of place. I call out as a gesture of faith: that country matters; that life has its own vibrancy, intensity, and modes of attention; and that my voice has a place in this world.*

And which Vaclav Havel talks about as well . . .

> *The kind of hope I often think about (especially in situations that are particularly hopeless, such as prison) I understand above all to be a state of mind, not a state of the world. Either we have hope within us or we don't; it is a dimension of the soul, and it's not essentially dependent on some particular observation of the world or estimate of the situation. Hope is not prognostication. It is an orientation of the spirit, an orientation of the heart; it transcends the world that is immediately experienced, and is anchored somewhere beyond its horizons. I don't think you can explain it as a mere derivative of something here, of some movement, or of some favorable signs in the*

world. I feel that its deepest roots are transcendental, just as the roots of human responsibility are. . . . An individual may affirm or deny that his hope is so rooted, but this does nothing to change my conviction (which is more than just a conviction; it is an inner experience). . . .

Hope, in this deep and powerful sense, is not the same as joy that things are going well, or the willingness to invest in enterprises that are obviously headed for early success, but, rather, an ability to work for something because it is good, not just because it stands a chance to succeed. The more unpropitious the situation in which we demonstrate hope, the deeper that hope is. Hope is definitely not the same thing as optimism. It is not the conviction that something will turn out well, but the certainty that something makes sense, regardless of how it turns out. In short, I think that the deepest and more important form of hope, the only one that can keep us above water and urge us to good works, and the only true source of the breathtaking dimension of the human spirit and its efforts, is something we get, as it were, from "elsewhere." It is also this hope, above all, which gives us the strength to live and continually try new things, even in conditions that seem as hopeless as ours do, here and now . . .

It helps me every day of my life to know these things. To hear these words. To know that people like Greta Thunberg and Deborah Bird Rose, Vaclav Havel and Masanobu Fuluoka, and so many others have walked among us. To know that we are not alone in this work, that all the kindred beings of this planet also work alongside us.

We are, as Bill Mollison said long ago, members of a nation that is unbounded by time or geography . . . or species. Every day of our lives we are surrounded by mystery, miracles of every sort, more companionship than we know, even when we think we are just living any old day at all.

Even now, even in the midst of all this loss and mourning, these things are true. Even now.

In veriditas veritas
The Gila Wilderness, March 2022.

Bibliography

Recommended texts and papers marked with an asterisk

TEXTS

Abram, David. *Becoming Animal: An Earthly Cosmology*, NY: Vintage, 2011.

Abram, David. *The Spell of the Sensous*, NY: Vintage, 1997.

Albrecht, Glenn. *Earth Emotions*, Ithaca, NY: Cornell University Press, 2019.

Arendt, Hannah. *Crises of the Republic*, NY: Harcourt Brace, 1969.

Bly, Robert, editor. *The Darkness Around Us Is Deep: Selected Poems of William Stafford*, NY: Harper Perennial, 1993.

Bly, Robert. *The Kabir Book*, Boston: Beacon Press, 1977.

• Bly, Robert. *A Little Book on the Human Shadow*, NY: Harper San Francisco, 1988.

Bly, Robert. *News of the Universe: Poems of Two Fold Consciousness*, San Francisco: Sierra Club Books, 1980.

Bly, Robert, James Hillman, Michael Meade. *The Rag and Bone Shop of the Heart*, NY: Harper Collins, 1992.

Brown, Valerie. Bacteria 'R' Us, *Pacific Standard Magazine*, December 2, 2010.

• Buhner, Stephen Harrod. *Healing Lyme*, second edition. Boulder, CO: Raven Press, 2015.

• Buhner, Stephen Harrod. *Herbal Antibiotics*, North Adams, MA: Storey Books, 2013.

• Buhner, Stephen Harrod. *Herbal Antivirals*, North Adams, MA: Storey Books, 2014.

• Buhner, Stephen Harrod. *The Lost Language of Plants*, White River Junction, VT: Chelsea Green, 2002.

• Buhner, Stephen Harrod. *Plant Intelligence and the Imaginal Realm*, Inner Traditions, 2014.

• Cunsolo, Ashlee and Karen Landman. *Mourning Nature: Hope at the Heart of Ecological Loss and Grief*, Montreal: McGill-Queen's University Press, 2017. Note: introduction only, the papers I found to be far too academic and dissociative.

Curry, Patrick. *Enchantment: Wonder in Modern Life*, Edinburgh: Floris Books, 2019.

Davis, Joseph and Paul Scherz. *The Evening of Life: The Challenges of Aging and Dying Well*, Notre Dame, Indiana: Notre Dame Press, 2020.

• Dreger, Alice. *Galileo's Middle Finger*, NY:Penguin, 2016.

Felman, Shoshana and Dori Laub. *Testimony*, NY: Routledge, 1992.

Gass, William. *Finding a Form*, NY: Knopf, 1996.

Gibson, Katherine and Deborah Bird Rose and Ruth Fincher, editors. *Manifesto for Living in the Anthropocene*, Brooklyn: Punctum Books, 2015. (no asterisk on this one)

Halifax, Joan and Stanislav Grof. *The Human Encounter with Death*, NY: Dutton, 1977.

Harries-Jones, Peter. *A Recursive Vision: Ecological Understanding and Gregory Bateson*, Toronto: University of Toronto Press, 1995.

Hatley, James. *Suffering Witness: The Quandary of Responsibility after the Irreparable*, Albany: State University of New York Press, 2000.

• Havel, Vaclav. *Disturbing the Peace*, NY: Knopf, 1990.

• Head, Lesley. *Hope and Grief in the Anthropocene*, London: Routledge, 2016.

Henderson, Keri and Joel Coats. *Veterinary Pharmaceuticals in the Environment*, Washington, D.C.: American Chemical Society, 2009.

Hester, R.E. and R.M. Harrison. *Pharmaceuticals in the Environment*, Cambridge, UK: The Royal Society of Chemistry, 2016.

Hillman, James. *A Blue Fire*, NY: Harper Perennial, 1989.

Hillman, James. *Re-visioning Psychology*, NY: Harper, 1977.

• Hillman, James. *Suicide and the Soul*, NY: Harper and Row, 1964.

Jamail, Dahr. *The End of ICE: Bearing Witness and Finding Meaning in the Path of Climate Disruption*, NY: The New Press, 2019.

Jensen, Tim. *Ecologies of Guilt in Environmental Rhetorics*, Switzerland: Palgrave Springer, 2019.

Jjemba, Patrick. *Pharma-ecology: The Occurrence and Fate of Pharmaceuticals and Personal Care products in the Environment*, second edition, Hoboken, NJ: John Wiley and Sons, 2019.

Kolbert, Elizabeth. *Under a White Sky: The Nature of the Future*, NY: Crown, 2021.

Kubler-Ross, Elizabeth. *To Live Until We Say Goodbye*, Englewood Cliffs, NJ: Prentice Hall, 1978.

Kubler-Ross, Elizabeth and David Kessler. *On Grief and Grieving*, NY: Scribner, 2005.

Kubler-Ross, Elizabeth. *Questions and Answers on Death and Dying*, NY: Macmillan, 1974.

Lappe Mark. *When Antibiotics Fail*, Berkeley, CA: North Atlantic Books, 1986.

• Leduc, Timothy. *A Canadian Climate of Mind*, Montreal: McGill-Queen's University Press, 2016.

• Leopold, Aldo. *A Sand County Almanac*, NY: Oxford University Press, 1989 (commemorative edition).

Levy Stuart, *The Antibiotic Paradox*, NY:Plenum Press, 1992.

Lewis, Michael. *Shame: The Exposed Self*, NY: The Free Press, 1995.

Margulis, Lynn and Dorion Sagan, *What Is Sex*, NY: Simon and Schuster, 1997.

Midgley, Mary. *The Essential Mary Midgley,* ed. David Midgley, London: Routledge, 2005. Note: All of her books I consider to be essential reading, she was one of the great thinkers of our time. This one gives the best overview of her work.

Midgley, Mary. *Evolution as Religion,* London: Routledge, 2002.

Midgley, Mary. *The Myths We Live By,* London: Routledge, 2004.

Midgley, Mary. *Science as Salvation,* London: Routledge, 1992.

Milosz, Czeslaw. *New and Collected Poems,* NY: Harper Collins, 2003.

Monbiot, George. *Feral: Rewilding the Land, the Sea, and Human Life,* Chicago: University of Chicago Press, 2014.

Norgaard, Kari Marie. *Living in Denial: Climate Change, Emotions, and Everyday Life,* Cambridge, MA: MIT Press, 2011.

Parkes, Colin and Holly G. Prigerson. *Bereavement: Studies of Grief in Adult Life,* fourth edition, NY: Penguin, 2010.

- Parkes, Colin and Robert S. Weiss. *Recovery From Bereavement,* NY: Basic Books, 1983.

- Plumwood, Val. *Environmental Culture: The Ecological Crisis of Reason,* London: Routledge, 2002. Note: A very good analysis of the limitations of reason when applied to ecological sustainability and its impacts on ecological health of the land.

- Riley, Denise. *Time Lived Without Its Flow,* London: Picador, 2012.

Rose, Deborah Bird. *Dingo Makes Us Human,* Cambridge, UK: Cambridge University Press, 2000.

- Rose, Deborah Bird. *Recursive Epistemologies and an Ethics of Attention,* in: Jean Goulet and Bruce Granville Miller (eds), *Extraordinary Anthropology,* Lincoln, NE: University of Nebraska Press, 2007.

Rose Deborah Bird. *Nourishing Terrains: Australian Aboriginal Views of Landscape and Wilderness,* Canberra: Australian Heritage Commission, 1996.

- Rose, Deborah Rose. *So the Future Can Come Forth from the Ground,* in: Kathleen Dean Moore and Michael Nelson (eds), *Moral Ground: Ethical Action for a Planet in Peril,* San Antonio, TX: Trinity University Press, 2010. Note: I found the book overall to be tremendously superficial, one of those where they get a lot of "names" to write a short piece. Rose's piece is, on the other hand, very good, although short.

Rose, Deborah Bird. *Wild Dog Dreaming,* Charlottesville, VA: University of Virginia Press, 2011.

Rose, Deborah Bird, Thom van Dooren, and Matthew Chrulew. *Extinction Studies,* NY: Columbia University Press, 2017.

- Scranton, Roy. *Learning to Die in the Anthropocene,* SF: City Lights Books, 2015.

Scranton, Roy. *We're Doomed, Now What?* NY: Soho, 2018.

Spellberg, Brad. *Rising Plague,* NY: Prometheus Books, 2009.

Stoknes, Per Espen. *What We Think About When We Try Not to Think About Global Warming,* White River Junction, VT: Chelsea Green, 2015.

Tarrant, John. *The Light Inside the Dark,* NY: Harper Collins, 1998.

Weller, Francis. *Entering the Healing Ground: Grief, Ritual, and the Soul of the World*, Santa Rosa, CA: WisdomBridge Press, 2011.

Weller, Francis. *The Wild Edge of Sorrow*, Berkeley, CA: North Atlantic Books, 2015.

Wolfelt, Alan. *The Wilderness of Grief*, Fort Collins, CO: Companion Press, 2007.

• Wood, Matthew. *Holistic Medicine and the Extracellular Matrix*, Inner Traditions, 2021.

• Wright, Ronald. *A Short History of Progress*, Philadelphia: Da Capo Press, 2004.

JOURNAL PAPERS AND MEDIA ARTICLES

Climate Crisis/Climate Grief

AFP in Paris, Climate crisis: dangerous thresholds to hit sooner than feared, UN report says, *The Guardian*, June 23, 2021.

• Ahmed, Nafeez. MIT predicted in 1972 that society will collapse this century. New research shows we're on schedule, *Vice*, July 14, 2021.

Ahmed, Nafeez. Theoretical Physicists Say 90% Chance of Societal Collapse Within Several Decades, *Vice*, July 28, 2020.

Alang, Navneet. Does Social Media Silence Grief, *The Week*, January 20, 2020.

Albrecht, G, et al. Solastalgia: the distress caused by environmental change, *Australas Psychiatry* 15 (supplement 1), 2007: S95-8.

Atkinson, Jennifer. Addressing climate grief makes you a badass, not a snowflake, *High Country News*, May 29, 2018.

Baker, Camille. Climate disasters will strain our mental health. It's time to adapt, *Washington Post*, September 4, 2021.

• Bardi, Ugo, et al. Toward a general theory of societal collapse: a biophysical examination of Tainter's model of the diminishing returns of complexity, *Biophysical Economics and Resource Quality* 4 (2019): 3.

Bendell, Jem. Deep Adaptation: A Map for Navigating Climate Tragedy, Institute for Leadership and Sustainability, Occasional Papers Volume 2, University of Cumbria, UK.

Berthold, Daniel. Aldo Leopold: In Search of a Poetic Science, *Research in Human Ecology* 11(3), 2004: 205-214.

Betts, Alan. What are our responsibilities to the Earth? alanbetts.com, downloaded January 21, 2020.

Bielinis, Ernest, et al. The effect of winter forest bathing on psychological relaxation of young Polish adults, *Urban Forestry and Urban Greening* 29, 2018: 276-83.

Bliss, Laura, The New Therapies for an Age of Climate Grief, *CityLab*, February 4, 2020.

• Boehnert, Jody Joanna. Epistemological Error and Converging Crises: A Whole Systems View, Learning from the Crisis of 2007-09, Philosophy of Management conference, May 2010, University of Oxford.

- Bortoft, Henri. The Transformative Potential of Paradox, *Holistic Science Journal* 1(1), nd.
- Bradshaw, Corey, et al. Underestimating the Challenges of Avoiding a Ghastly Future, *Frontiers in Conservation Science* 1, 2021, article 615419.
- Brannon, Peter. The Terrifying Warning Lurking in the Earth's Ancient Rock Record, *The Atlantic*, February 3, 2021.
- Bromwich, Jonah. The Darkest Timeline, *New York Times*, December 26, 2020.
- Buckley, Cara. Apocalypse got you down? Maybe this will help, *New York Times*, November 15, 2019.
- Cafula, Anna. A Future World: Climate Change and Mental Health, the People Grieving for the Planet, *Dazed*, May 15, 2019.
- Carrington, Damian. "Insect apocolypse" poses risk to all life on Earth, conservationists warn, *The Guardian*, November 13, 2019.
- Castagnetti, Francesca. The Language of the Land, *The Ethnobotanical Assembly*, June 14, 2019.
- Cavanagh, Michaela. It's Time to Talk About Ecological Grief, *Undark*, January 10, 2019.
- Chang, Connie. How to cope with the existential dread of climate change, *Washington Post*, July 15, 2021.
- Chapman, Sasha. Playing Chicken, *The Walrus*, August 23, 2017.
- Cooperrider, Kensy. What happens to cognitive diversity when everyone is more WEIRD? *Aeon*, January 23, 2019.
- Corn, David. It's the End of the World as They Know It, *Mother Jones*, July 8, 2019.
- Cunsolo, Ashlee and Neville Ellis. Ecological grief as a mental health response to climate change-related loss, *Nature Climate Change* 8, 2018: 275-81.
- Dai, Lei, et al. Generic indicators for loss of resilience before a tipping point leading to population collapse, *Science* 336 (2012): 1175-7.
- Dechristopher, Tim. To Live and Love with a Dying World, *Orion Magazine*, downloaded November 1, 2020.
- Deighton, Russell and Norbert Gurris and Harold Traue. Factors Affecting Burnout and Compassoin Fatigue in Psychotherapists Treating Torture Survivors: Is the Therapist's Attitudes to Working Through Trauma Relevant? *Journal for Traumatic Stress* 20(1), 2007: 63-75.
- Deresiewicz, William. The End of Solitude, *Hermitary*, downloaded December 23, 2019.
- Dibley, Ben. The Shape of Things to Come: Seven Theses on the Anthropocene and Attachment, *Australian Humanities Review*, Issue 52.
- Doherty, Thomas and Susan Clayton. The Psychological Impacts of Global Climate Change, *American Psychologist* 66(4), 2011: 265-76.
- Douillard, John. Fear changes soil chemistry – what's it doing to us? *Rare*, December 1, 2018.

Dworkin, Ronald. The Limits of Science, *National Affairs*, Winter 2019.

Ellis, Neville and Ashlee Cunsolo, Hope and mourning in the Anthropocene – understanding ecological grief, *The Conversation*, April 5, 2018.

Emmott, Stephen. Humans: the real threat to life on Earth, *The Guardian*, June 29, 2013.

• Flyn, Cal. The Best of Nature Writing 2019, Five Books, December 15, 2019.

Francis, Ellen. Activists "born in the climate crisis" face another challenge: Fear of the future, *Washington Post*, September 16, 2021.

Frank, Adam and Maecelo Gleiser and Evan Thompson. The Blind Spot, *Aeon*, downloaded February 13, 2019.

Fritze, Jessica, et al. Hope, despair and transformation: Climate change and the promotion of mental health and wellbeing, *International Journal of Mental Health Systems* 2(13), 2008.

Frost, Matt. After Climate Despair, *The New Atlantis*, downloaded December 27, 2020.

Gammon, Katherine. Critical measures of global healing reaching tipping point, study finds, *The Guardian*, July 27, 2021.

Garber, Megan. The Dark Side of the Houseplant Boom, *The Atlantic*, April 20, 2021.

Garfinkle, Adam. The Erosion of Deep Literacy, *National Affairs*, Spring 2020.

• Gerson, Samuel. When the Third Is Dead: Memory, Mourning, and Witnessing in the Aftermath of the Holocaust, *The International Journal of Psychoanalysis* 90, 2009: 1341-57.

Ghimire, Poonam. My Earth, My Responsibility, *Voices of Youth*, November 20, 2014.

Gibbens, Sarah. Protecting land and animals will mitigate future pandemics, report says, *National Geographic*, October 29, 2020.

Gould, Catie. How I learned to cope with climate grief, *BikePortland*, March 5, 2020.

Goulson, Dave. The insect apocalypse: Our world will grind to a halt without them, *The Guardian*, July 25, 2021.

Gregory, Alice. The Sorrow and the Shame of the Accidental Killer, *The New Yorker*, September 18, 2017.

Grose, Anouchka. How the climate emergency could lead to a mental health crisis, *The Guardian*, August 13, 2019.

Gwilliam, Jesse. Tossing climate guilt in the gargage, *Massachusetts Daily Collegian*, October 1, 2019.

Hansen, Margaret and Reo Jones and Kirsten Tocchini. Shinrin-Yoku (Forest Bathing) and Nature Therapy: A State-of-the-Art Review, *International Journal of Environmental Research and Public Health* 14(8), 2017.

Harrison, Melissa. Feeling severely distressed about the climate crisis? You're suffering from solastalgia, *New Statesman*, October 23, 2019.

Harvey, Fiona. Major climate changes inevitable and irreversible – IPCC's starkest warning yet, *The Guardian*, August 9, 2021.

Harvey, Fiona. Reduce methane or face climate catastrophe, scientists warn, *The Guardian*, August 6, 2021.

Hatley, James. The Virtue of Temporal Discernment: Rethinking the Extent and Coherence of the Good in a Time of Mass Species Extinction, *Environmental Philosophy* 9(1), 2012: 1-22.

- Head, Lesley. The Anthropoceneans, *Geographical Research* 53, 2015: 313-20.

Head, Lesley. Contingencies of the Anthropocene: Lessons from the Neolithic, *The Anthropocene Review*, 2014: 1-13.

Head, Lesley. Grief, Loss, and the Cultural Politics of Climate Change, Columbia University Libraries, [www]cambridge.org/core, downloaded August 16, 2017.

- Heglar, Mary Annaise. I work in the environmental movement. I don't care if you recycle, *Vox*, June 4, 2019.

- Herrington, Gaya. Update to limits to growth: Comparing the World3 model with empirical data, *Journal of Industrial Ecology*, 2020: 1-13.

Hill, Samatha Rose. When hope is a hindrance, Aeon, October 10, 2021.

Hochuli, Alex. The Brazilianization of the World, *American Affairs*, 2021.

Jacquet, Jennifer. Human Error: Survivor guilt in the Anthropocene, *Lapham's Quarterly*, downloaded January 21, 2020.

- Jamail, Dahr. In Facing Mass Extinction, We Must Allow Ourselves to Grieve, *Truthout*, January 17, 2019.

Jones, Lucy. Ecological grief; I mourn the loss of nature – it saved me from addiction, *The Guardian*, February 25, 2020.

Kaplan, Sarah. Many measures of Earth's health are at worst levels on record, NOAA finds, *Washington Post*, August 26, 2021.

Keesee, Nancy and Joseph Currier and Robert Neimeyer. Predictors of Grief Following the Death of One's Child: The Contribution of Finding Meaning, *Journal of Clinical Psychology* 64(10), 2008: 1145-63.

Kendrick, Gary, et al. A systematic review of how multiple stressors from an extreme event drove ecosystem-wide loss of resilience in an iconic seagrass community, *Frontiers in Marine Science* 6 (455), 2019.

Kingsnorth, Paul. Dark Ecology, *Orion Magazine*, December 21, 2012.

Kingwell, Mark. How to Live with Death, *The Walrus*, October 14, 2020.

Klinkenborg, Verlyn. How the Loss of Soil is Sacrificing America's Natural Heritage, *Yale Environment*, March 1, 2021.

Knight, Victoria. Climate Grief: Fears About the Planet's Future Weigh on Americans' Mental Health, *Kaiser Health News*, July 18, 2019.

- Kreitler, Shulamith, et al. Survivor's Guilt in Caretakers of Cancer, in: Kate Hinerman and Julia Glahn, eds. The Presence of the Dead in Our Lives, *At the Interface/Probing the Boundaries* 82, 2012.

Langlitz, Nicolas. Salvage and self-loathing: Cultural primatology and the spiritual malaise of the Anthropocene, *Anthropology Today* 34(16), 2018: 16-20.

Larson, Erik. When Making Things Better Only Makes Them Worse, *The Atlantic*, April 26, 2019.

Law, Rob. I have felt hopelessness over climate change. Here is how we move past the immense grief, *The Guardian*, May 8, 2019.

- LeBrun, Annie. Priceless: Beauty, Ugliness, and Politics, sample material translated (from the French) by John Galbraith Simmons, *Hyperion*, 2018.
- Leduc, Timothy. Falling with Heron: Kaswen:ta teachings on roughening waters, *Social and Cultural Geography*, 2018. DOI:10.1080/14649365.2018.1500633.

Lee, J, et al. Effect of forest bathing on physiological and psychological responses in young Japanese male subjects, *Public Health* 125(2), 2011: 93-100.

Lenton, Timothy, et al. Climate tipping points – too risky to bet against, *Nature* 575, November 28, 2019.

Li, Qing. Effect of forest bathing trips on human immune function, *Environmental Health and Preventative Medicine* 15, 2010: 9-17.

Li, Q et al. Forest Bathing Enhances Human Natural Killer Activity and Expression of Anti-cancer Proteins, *International Journal of Immunopathology and Pharmacology* 20(2), 2007: 3-8.

Liberatore, Stacy. Declining resilience of North America's plant biomes may be a sign of a mass extinction last seen nearly 13,000 years ago, experts warn, *Daily Mail*, August 31, 2020.

Macdonald, Fiona. The Vanishing Words We Need to Save, BBC Culture, November 26, 2015.

Magnason, Andri Snaer. What Will Be the Tipping Point of the Climate Crisis? *The Walrus*, March 29, 2021.

Malloy, Michael. Montana's Money Pit, *The Baffler*, January 2020.

Mao, Gen-Xing, et al. The Salutary Influence of Forest Bathing on Elder Patients With Chronic Heart Failure, *International Journal of Environmental Research and Public Health* 14(4), 2017.

Mao, Gen-Xing, et al. Therapeutic effect of forest bathing on human hypertension in the elderly, *Journal of Cardiology* 60(6), 2012: 495-502.

- Marchese, David. Greta Thunberg Hears Your Excuses. She Is Not Impressed, *New York Times*, October 30, 2020.

McDonald, Samuel. The Ministry for the Future, or Do Authors Dream of Electric Jeeps? *Current Affairs*, January 24, 2021.

McKenna, Maryn. The Catch-22 of Mass-Prescribing Antibiotics, *Wired*, May 10, 2018.

Meyer, Robinson. It's Grim, *The Atlantic*, August 9, 2021.

Miner, Kimberley and Arwyn Edwards and Charles Miller. Deep Frozen Microbes Are Waking Up, *Scientific American*, November 20, 2020.

Molteni, Megan. The Post-antibiotic Era is Here. Now What? *Wired*, September 25, 2017.

Monbiot, George. Earth's tipping points could be closer than we thing: Our current plans won't work, *The Guardian*, September 9, 2021.

Monbiot, George. Trashing the planet and hiding the money isn't a perversion of capitalism. It is capitalism. *The Guardian*, October 6, 2021.

Moore, Suzanne. In grief, I have found unexpected comforts, *The Guardian*, October 29, 2019.

Mull, Amanda. The Difference Between Being Safe and Feeling Safe, *The Atlantic*, October 26, 2020.

Murphy, Layla. Finding My Will to Fight for the Planet, *34th Street Magazine*, November 19, 2019.

Naipaul, V.S. The Strangeness of Grief, *The New Yorker*, December 30, 2019.

Newhouse, Alana. Everything Is Broken, *Tablet Magazine*, January 14, 2021.

Nightshade. Earth-Grief: The World Soul, and Stories of Sorrow, Grief and Healing, *The Purple Broom*, purplebroom.com, downloaded December 15, 2019.

O'Rourke, Ciara. Climate Change's Hidden Victim: Your Mental Health, *Medium*, January 24, 2019.

Ortiz, Diego. Real hope for the future of the climate can't come from admiring the inspiration deeds of others – it has to be earned, BBC Future, January 9, 2020.

Osaka, Shannon. 'The planet is broken,' U.N. chief says, *Grist*, December 3, 2020.

Park, Bum Jin, et al. The physiological effects of shinrin-yoku (taking in the forest atmosphere or forest bathing): evidence from field experiences in 24 forests across Japan, *Environmental Health and Preventative Medicine* 15, article 18, 2010.

Paul, Norman. The Use of Empathy in the Resolution of Grief, *Perspectives in Biology and Medicine*, Autumn 1967.

- Pauly, Daniel. Anecdotes and the shifting baseline syndrome of fisheries, *TREE* 10(10), 1995.

Pearce, Fred. Water Warning: The Looming Threat of the World's Aging Dams, *Yale Environment*, February 3, 2021.

Pearl, Mike. Climate Despair Is Making People Give Up on Life, *Vice*, July 11, 2019.

Pearson, Patricia. The Danger of Putting Youth on Antidepressants, *The Walrus*, September 20, 2017.

Pierrehumbert, Ray and Michael Mann. Some say we can 'solar-engineer' ourselves out of the climate crisis. Don't buy it, *The Guardian*, April 22, 2021.

Pihkala, Panu. Climate Grief: How we mourn a changing planet, BBC.com, April 2, 2020.

- Plumwood, Val. Being Prey, *Terra Nova* 1(3), 1996.

- Plumwood, Val. Tasteless: Towards a Food-based Approach to Death, *Environmental Values* 17, 2008: 323-30.

Pontecorvo, Emily. Abandonment Issues: The Number of Abandoned Oil and Gas Wells Is on the Verge of Exploding, *Grist*, December 1, 2020.

Prendergast, Conor. Solar Panel Waste: The Dark Side of Clean Energy, *Discover Magazine*, December 14, 2020.

Preston, Caroline. Depressed about climate change? There's a 9-step program for that, *Grist*, April 8, 2017.

Ramamurthy, Rithika. Personal Hell: the Climate Anxiety Novel, *The Drift*, May 6, 2021.

Randall, Rosemary. Loss and Climate Change: The Cost of Parallel Narratives, *Ecopsychology* 1(3), September 2009.

Ripple, William, et al. World Scientists' Warning of a Climate Emergency, *Bioscience* 70(1), 2020.

Robbins, Jim. Ecopsychology: How Immersion in Nature Benefits Your Health, *Yale Environment 360*, January 9, 2020.

Robbins, Paul and Sarah Moore. Ecological anxiety disorder: Diagnosis the politics of the Anthropocene, *Cultural Geographies* 20(1), 2012: 3-19.

• Rose, Deborah Bird. *Anthropocene Noir*, Global Cities Research Institute, 2013.

• Rose, Deborah Bird. Ruined Faces, in *Facing Nature*, William Edelglass, James Hatley, and Christian Diehm (eds), Pittsburg: Duquessne University Press, 2012.

Rose, Deborah Bird. Judas Work: Four Modes of Sorrow, *Environmental Philosophy* 5(2), 2008: 51-66.

• Rose, Deborah Rose. In the Shadow of All This Death, in *Animal Death*, Jay Johnston and Fiona Probyn-Rapsey, Sydney: Sydney University Press, 2013.

Rose, Deborah Rose. Love in the Time of Extinctions, *The Australian Journal of Anthropology*.

• Rose, Deborah Bird. Multispecies Knots of Ethical Time, *Environmental Philosophy* 9(1), 2012: 127-40.

• Rose, Deborah Bird. On history, trees, and ethical proximity, *Postcolonial Studies* 11(2), 2008: 157-67.

• Rose, Deborah Bird. Why I Don't Speak of Wilderness, *EarthSong Journal: Perspectives in Ecology, Spirituality and Education*, September 1, 2012.

Rosenfeld, Jordan. Facing Down "Environmental Grief," *Scientific American*, July 21, 2016.

Running, Steven. The 5 Stages of Climate Grief, University of Missoula, 2007.

Safina, Carl. Avoiding a 'Ghastly Future': Hard Truths on the State of the Planet, *Yale Environment 360*, January 27, 2021.

Safina, Carl. Psychic Numbing: Keeping Hope Alive in a World of Extinctions, *Yale Environment 360*, February 26, 2020.

St. George, Zach. As Climate Warms, a Rearrangement of World's Plant Life Looms, *Yale Environment 360*, June 17, 2021.

• Sanders, Ash. Under the Weather, *The Believer*, December 2, 2019.

Scher, Avichai. Climate grief hits the self-care generation, *CUNY Academic Works*, Fall 2018.

Schwagerl, Christian. What's Causing the Sharp Decline in Insects and Why It Matters, *Yale Environment 360*, July 6, 2016.

Scialabba, George. Back to the Land, *The Baffler*, number 49, January 2020.

Scranton, Roy. I've Said Goodbye to 'Normal.' You Should, Too. *New York Times*, January 25, 2021.

Scranton, Roy. Learning How to Die in the Anthropocene, *New York Times*, November 10, 2013.

Shah, Sonia. Native Species or Invasive? The Distinction Blurs as the World Warms, *Yale Environment 360*, January 14, 2020.

Shain, Susan. Got Climate Anxiety? These People Are Doing Something About It, *New York Times*, February 4, 2021.

Shaw, Wendy and Alastair Bonnett. Environmental crisis, narcissism and the work of grief, *Cultural Geographies*, 2016:1-15.

Sheng-Yang He. When plants and their microbes are not in sync, the results can be disastrous, *The Conversation*, August 28, 2020.

Siegel, Lee. Whatever Happened to Moral Rigor? *The New York Times*, July 25, 2018.

Spady, Donald. The environment and our responsibility to our children and youth: a message for adults, *Pediatrics and Children's Health* 14(5), 2009.

Steffen, Will, et al. Trajectories of the Earth System in the Anthropocene, *PNAS* 115(33), 2018: 8252-59.

Stoknes, Per Espen. The Great Grief: How to Cope with Losing Our World, *Common Dreams*, May, 14, 2015.

Taibbi, Matt. Interview with Martin Gurri, "A Short-Term Pessimist and a Long-Term Optimist," *TK News (substack)*, March 8, 2021.

Tampio, Nicholas. Look up from your screen, *Aeon*, August 2, 2018.

Taylor, Matthew and Jessica Murray. Overwhelming and terrifying: the rise of climate anxiety, *The Guardian*, February 10, 2020.

- Thunberg, Greta, Adriana Calderon, Farzana Faruk Jhumu, and Eric Njuguna. This is the world being left to us by adults, *New York Times*, August 19, 2021.

Torino, Giulia. Age of Extraction: An Interview with Saskia Sassen, *Kings Review*, November 24, 2017.

Turner, Chris. We're Doomed. Now What? *The Walrus*, November 4, 2019.

Vallance, Patrick. The IPCC report is clear: nothing short of transforming society will avert catastrophe, *The Guardian*, August 9, 2021.

Voosen, Paul. U.N. climate panel confronts implausible hot forecasts of future warming, *Science*, July 27, 2021.

Watts, Jonathan. Vaclav Smil: Growth must end. Our economist friends don't seem to realize that, *The Guardian*, September 21, 2019.

Weil, Elizabeth. The Climate Crisis Is Worse Than You Can Imagine. Here's What Happens If You Try, *Tablet Magazine*, January 25, 2021.

Weschler, Lawrence. Beyond Climate Denial and Despair, *The Atlantic*, December 15, 2020.

Whang, Oliver. Greta Thunberg reflects on living through multiple crises in a "post truth society," *National Geographic*, October 28, 2020.

White, Kristi. Climate Change Is the New Disease of Despair. Psychologists Need to Step Up to Help, *Ensia*, February 21, 2020.

Whiting, Tabitha. You Need to Stop Feeling Guilty about Climate Change, *Medium*, July 20, 2019.

- Williams, Rowan. How Dying Offers a Chance to Live the Fullest Life, *New Statesman*, April 15, 2018.

Willox, Ashlee Cunsolo. Climate Change as the Work of Mourning, *Ethics and the Environment* 17(2), 2012: 137-164.

- Windle, Phyllis. The Ecology of Grief, *BioScience* 42(5), 1992.

Witztum, Eliezer and Ruth Malkinson. Examining Traumatic Grief and Loss Among Holocaust Survivors, *Journal of Loss and Trauma* 14, 2009: 129-43.

Wolchover, Natalie. Nature's Warning Signal, *The Atlantic*, December, 2015.

World Staff. Is climate change causing us to experience "ecological grief"? *The World*, June 24, 2019.

Yoder, Kate. Want some eco-friendly tips? A new study says no, you don't, *Grist*, October 12, 2020.

Zenith, Shante' Sojourn. Grief-Tending and the Ecological Imagination, *Transition United States*, July 2, 2018.

Plastics:

Alnajar, Nashami. Impacts of microplastic fibres on the marine mussel, *Mytilus galloprovinciallis, Chemosphere* 262, 2021: 128290.

Andrews, Luke. Microplastics are contaminating the fruit and vegetables we eat including apples, carrots and lettuces after being absorbed through their roots, studies show, *Daily Mail* online, June 25, 2020.

Auguste, Manon et al. Impact of nanoplastics on hemolymph immune parameters and microbiota composition in Mytilus galloprovincialis, *Marine Environmental Research* 159, 2020:

105017.

Bourn, Chris. Can the construction industry solve the world's plastic crisis? *Melmagazine*, September 9, 2020.

Browne, Grace. Why food's plastic problem is bigger than we realize, BBC online, downloaded February 3, 2020.

Carrington, Damian. Air pollution particles found on foetal side of placentas – study, *The Guardian*, September 17, 2019.

Carrington, Damian. Car tyres are major source of ocean microplastics – study, *The Guardian*, July 14, 2020.

Carrington, Damian. Clothes washing linked to 'pervasive' plastic pollution in the arctic, *The Guardian*, January 12, 2021.

Carrington, Damian. Microplastic particles now discovered in human organs, *The Guardian*, August 17, 2020.

Carrington, Damian. Microplastics "signifcantly contaminating the air," scientists warn, *The Guardian*, August 14, 2019.

Catro, Joseph. Plastic Legacy: Humankind's Trash Is Now a New Rock, *Live Science*, June 3, 2014.

CBC Docs. Recycling was a lie – a big lie – to sell more plastic, industry experts say, *CBC Documentaries*, September 23, 2020.

Dickinson, Tim. Planet plastic: how big oil and big soda kept a global environmental calamity a secret for decades, *Rolling Stone*, March 3, 2020.

Ding, Haojie, et al. Do membrane filtration systems in drinking water treatment plants release nano/microplastics? *Science of the Total Environment* 755 (part 2), 2021: 142658.

E360 Digest. As Plastic Pollution in Rivers Gets Worse, Species Are Increasingly Living on Litter, *Yale Environment 360*, February 9, 2021.

Farrier, David. Hand in Glove, *Orion Magazine*, downloaded September 11, 2020.

Fernandez, Colin. Microplastic particles in the womb: women give birth to 'cyborg babies' no longer made up of just human cells as experts fear it could interfere with development, *Daily Mail* online, December 18, 2020.

Fournier, Sara, et al. Nanopolystyrene translocation and fetal deposition after acute lung exposure during late-stage pregnancy, *Particle and Fibre Toxicology* 17(1), 2020: 55.

Gangadoo, S, et al. Nano-plastics and their analytical characterization and fate in the marine environment: from source to sea, *Science of the Total Environment* 732, 2020: 138792.

Gardiner, Beth. The Plastics Pipeline: A Surge of New Production Is on the Way, *Yale Environment 360*, December 19, 2019.

Gu, Huaxin, et al. Nanoplastics impair the intestinal health of the juvenile large yellow croaker Larimichthys crocea, *Journal of Hazardous Materials* 397, 2020: 122773.

Harvey, Fiona. Atlantic ocean plastic more than 10 times previous estimates, *The Guardian*, August 18, 2020.

Hirt, Neil and Mathilde Body-Malapel. Immunotoxicity and intestinal effects of nano- and microplastics: a review of the literature, *Particle and Fibre Toxicology* 17, 2020: 57.

Horton, Adrian. John Oliver on plastics pollution: "Our personal behavior is not the main culprit," *The Guardian*, March 22, 2021.

Huang, Zhuizui, et al. Microplastic: a potential threat to human and animal health by interfering with the intestinal barrier function and changing the intestinal microenvironment, *Science of the Total Environment* 785, 2021: 147365.

Jung, Byung-Kwon et al. Neurotoxic potential of polystyrene nanoplastics in primary cells originating from mouse brain, *Neurotoxicology* 81, 2020: 189-96.

Kane, Ian, et al. Seafloor microplastic hotspots controlled by deep-sea circulation, *Science* 10.1126/science.aga5899, 2020.

Lerner, Sharon. Africa's Exploding Plastic Nightmare, *The Intercept*, April 19, 2020.

Levin, Myron. Tenacious citizens take on the plastics industry over an insidious pollutant, *Salon*, October 3, 2020.

Lew, Joseph. Oh great, scientists are now finding traces of plastic in human flesh, *Vice*, August 25, 2020.

Li, Yiming et al. Effects of nanoplastics on antioxidant and immune enzyme activities and related gene expression in juvenile *Macrobrachium nipponense*, *Journal of Hazardous Materials* 398, 2020: 122990.

Lu, Liang, et al. Interaction between microplastics and microorganism as well as gut microbiota: a consideration on environmental animal and human health, *Science of the Total Environment* 667, 2019: 94-100.

Lungstrom, Marjie and Eli Wolfe. Fields of Waste: Artificial Turf, Touted as Recycling Fix for Millions of Scrap Tires, Becomes Mounting Disposal Mess, *FairWarning*, December 19, 2019.

Machado, A, et al. Microplastics as an emering threat to terrestrial ecosystems, *Global Change Biology* 24(4), 2018: 1405-16.

Milman, Oliver. Pollution from car tires is killing off salmon in US west coast, study finds, *The Guardian*, December 3, 2020.

Ngo, Hope. How do you fix healthcare's medical waste problem? BBC.com, August 13, 2020.

Pabortsava, Karsiaryna and Richard Lampitt. High concentrations of plastic hidden beneath the surface of the Atlantic Ocean, *Nature Communications* 11, 2020: 4073.

Parker, Laura. Microplastics have moved into virtually every crevice on Earth, *National Geographic*, August 7, 2020.

Prust, M et al. The plastic brain: neurotoxicity of micro- and nanoplastics, *Particle and Fiber Toxicology* 17, 2020: 24.

Ramsperger, A, et al. Environmental exposure enhances the internalization of microplastic particles into cells, *Science Advances* 6, 2020: eabd1211.

Randall, Ian. Plants can absorb tiny pieces of plastic through their roots that stunt their growth and reduce their nutritional value, study shows, *Daily Mail* online, June 23, 2020.

Randall, Ian. Scientists find plastic in sea turtles' muscles for the first time, confirming that pollution is affecting marine life on a chemical level, *Daily Mail* online, May 5, 2021.

Robbins, Jim. Why Bioplastics Will Not Solve the World's Plastics Problem, *Yale Environment 360*, August 31, 2020.

Rozsa, Matthew. How plastics are making us infertile – and could even lead to human extinction, *Salon*, April 4, 2021.

Rozsa, Matthew. What is microplastic anyway? Inside the insidious pollution that is absolutely everywhere, *Salon*, July 17, 2021.

Rubio, Laura, et al. Biological effects, including oxidative stress ad genotoxic damage, of polystyrene nanoparticles in different human hematopoietic cell lines, *Journal of Hazardous Materials* 398, 2020: 122900.

Simon, Matt. Plastic Rain Is the New Acid Rain, *Wired*, June 11, 2020.

Smedley, Tim. Current pollution meters don't count the very smallest pollutants – nanoparticles. Recent research suggests these tiny toxic substances could be a major cause of illness and death, BBC Future, November 15, 2019.

Stuart, Ryan. Scooping plastic out of the ocean is a losing game, *Hakai Magazine*, September 21, 2021.

Thomas, Russell. Welcome to the "plastisphere": the synthetic ecosystem evolving at sea, *The Guardian*, August 11, 2021.

Tian, Zhenyu, et al. A ubiquitous tire rubber-derived chemical induces acute mortality in coho salmon, *Science* (Reports) 10.1126/science.abd6951, 2020.

Tierney, John. The Reign of Recycling, *New York Times*, October 3, 2015.

Tobuchi, Hiroko and Michael Corkery and Carlos Mureithi, Big Oil Is in Trouble. Its Plan: Flood Africa with Plastic, *New York Times*, August 30, 2020.

Tong, Ziya. We Are Garbage, *The Walrus*, August 26, 2019.

Unkown. The Great Nurdle Hunt, nurdlehunt.org.uk, downloaded October 20, 2020.

Watts, Jonathan. Microplastic pollution devastating soil species, study finds, *The Guardian*, September 1, 2020.

Weisman, Alan. Polymers Are Forever, *Orion Magazine*, downloaded September 11, 2020.

Woollacott, Emma. Our plastic waste is changing the geology of the Earth's rocks, *New Statesman*, June, 2014.

Zanolli, Lauren and Mark Oliver. Explained: the toxic threat in everyday products, from toys to plastics, *The Guardian*, May 22, 2019.

Pharmaceuticals and the Medical Industry:

Agence France-Presse. Drug waste clogs rivers around the world, scientists say, *The Guardian*, April 10, 2018.

American Rivers. Pharmaceuticals in the Water Supply, *American Rivers*, downloaded October 8, 2019.

Arnold, Kathryn, et al. Assessing the exposure risk and impacts of pharmaceuticals in the environment on individuals and ecosystems, *Biology Letters* 9, 2013: 0492.

Bain, Kevin. Public health implications of household pharmaceutical waste in the United States, *Health Services Insights* 3, 2010: 21-36.

Beek, Tim aus der, et al. Pharmaceuticals in the environment – global occurrences and perspectives, *Environmental Toxicology and Chemistry* 35(4), 2016: 823-35.

Biello, David. Pill to Gill: Antianxiety Drugs Flushed into Water May Be Making Fishes Fearless, *Scientific American*, February 15, 2013.

Brodin, Tomas, et al. Environmental relevant levels of a benzodiazepine (oxazepam) alters important behavioral traits in a common planktivorous fish (*Rutilus rutilus*), *Journal of Toxicology and Environmental Health, Part A*, 2017: 1352214.

Burnham, Jason. Re-estimating annual deaths due to multidrug-resistant organism infections, *Infection Control and Hospital Epidemiology* 40(1), 2019:112-3.

Chapman, Sasha. Playing Chicken, *The Walrus*, January 16, 2015.

Chung, Shan-shan and Bryan Brooks. Identifying household pharmaceutical waste characteristics and population behaviors in one of the most densely populated global cities, *Resources, Conservation & Recycling* 140, 2019:267-77.

Collin, Johanne. On social plasticity: the transformative power of pharmaceuticals on health, nature, and identity, *Sociology of Health & Illness* 38(1), 2016: 73-89.

Davies, Madlen. Big pharma's pollution is creating deadly superbugs while the world looks the other way, *Bureau of Investigative Journalism*, May 6, 2017.

Deeb, Ahmad, et al. Suspect screening of micropollutants and their transformation products in advanced wastewater treatment, *Science of the Total Environment* 601-602, 2017: 1247-53.

Dellinger, AJ. Coronavirus is creating a staggering amount of medical waste, mic.com, March 27, 2020.

El Murr, Yara. Hospitals try to curb astronomical emissions as pandemic brings new challenges, *The Guardian*, April 7, 2021.

Evans, Sydney, et al. PFAS contamination of drinking water far more prevalent than previously reported, ewg.org, January 22, 2020.

Fletcher, Carly. What happens to waste PPE during the coronavirus pandemic? *The Conversation*, May 12, 2020.

Galdiero, F, et al. Effects of benzodiazepines on immunodeficiency and resistance in mice, *Life Sciences* 57(26), 1995: 2413-23.

• Gerber, Leah. This little-known principle has harmed millions of people. What are we doing to change it? *Ensia*, downloaded April 5, 2020.

Giggs, Rebecca. Human Drugs Are Pollution the Water – And Animals Are Swimming in It, *The Atlantic*, May 2019.

Gilbert, Natasha. Dump it down the drain, *Type Investigations*, December 11, 2019.

Gilbert, Natasha. World's rivers "awash with dangerous levels of antibiotics," *The Guardian*, May 26, 2019.

Guillette, Louis. Endocrine disrupting contaminants – beyond the dogma, *Environmental Health Perspectives* 114 (supplement 1), 2006:9-12.

Guillette, Louis, et al. Alligators and Endocrine Disrupting Contaminants: A Current Perspective, *American Zoology* 40, 2000:438-52.

Guillette, Louis, et al. Epigenetic programming alterations in alligators from environmentally contaminated lakes, *General and Comparative Endocrinology* 238, 2016: 4-12.

Guthrie, George and Catherine Nicholson-Guthrie. y-Aminobutyric acid uptake by a bacterial system with neurotransmitter binding characteristics, *Proceedings of the National Academy of Sciences* 86, 1989: 7378-81.

Hess, Jeremy, et al. Petroleum and Health Care: Evaluating and Managing Health Care's Vulnerability to Petroleum Supply Shifts, *Peak Petroleum and Public Health* 101(9), 2011:1568-79.

Hughes, Stephen and Paul Kay and Lee Brown. Global synthesis and critical evaluation of pharmaceutical data sets collected from river systems, *Environmental Science and Technology* 47, 2013: 661-77.

Jaseem, Muhammed, et al. An overview of waste management in pharmaceutical industry, *Pharma Innovation Journal* 6(3), 2017: 158-61.

Jobling, Susan, et al. Predicted exposures to steroid estrogens in U.K. rivers correlate with widespread sexual disruption in wild fish populations, *Environmental Health Perspectives* 114 (supplement 1), 2006.

Joseph, Brian. Study links cosmetics to altered body chemistry, *Fair Warning Website*, March 7, 2016.

Kamba, Pakoyo, et al. Why regulatory indifference towards pharmaceutical pollution of the environment could be a missed opportunity in public health protection, a holistic view, *PanAfrican Medical Journal* 27, 2017: 77.

Kolpin, Dana, et al. Pharmaceuticals, hormones, and other organic wastewater contaminants in U.S. streams, 1999-2000: a national reconnaissance, *Environmental Science and Technology* 36, 2002:1202-11.

Kookana, Rai, et al. Potential ecological footprints of active pharmaceutical ingredients: an examination of risk factors in low-, middle-, and high-income countries, *Philosophical Transactions of the Royal Society B* 369, 2013: 0586.

Kuster Anette and Nicole Adler. Pharmaceuticals in the environment: scientific evidence of risk and its regulation, *Philosophical Transactions of the Royal Society B* 369: 20130587.

Kosjek, T. et al. Environmental occurrence, fate and transformation of benzodiazepines in water treatment, *Water Research* 46, 2012: 355-68.

Kummerer, Klaus. Pharmaceuticals in the Environment, *Annual Review of Environment and Resources* 35, 2010: 57-75.

Kurtzman, Laura. UCSF Study Finds Evidence of 55 Chemicals Never Before Reported in People, *University of California San Francisco Research*, March 16, 2021.

Law, Anandi, et al. Taking stock of medication wastage: unused medications in US households, *Research in Social and Administrative Pharmacy*, 2014:1-8.

Lim, Xiao Zhi. Tainted water: the scientists tracing thousands of fluorinated chemicals in our environment, *Nature*, February 6, 2019.

Lohan, Tara. What happens to wildlife swimming in a sea of our drug residues? *Salon*, August 17, 2021.

Lubbert, Christoph, et al. Environmental pollution with antimicrobial agents from bulk drug manufacturing industries in Hyderabad, South India, is associated with dissemination of extended-spectrum beta-lactamase and carbapenemase-producing pathogens, *Infection* online, 2017.

Maynard, Jake. Green Burial Wants to Clean Up American Funerals, *Slate*, February 5, 2021.

McKenna, Maryn. Racing the clock to stop drug-resistant superbugs, *Boston Globe*, August 5, 2020.

Miller, Thomas, et al. Biomonitoring of pesticides, pharmaceuticals and illicit drugs in a freshwater invertebrate to estimate toxic or effect pressure, *Environment International* 129, 2019: 595-606.

Miller, Thomas, et al. A review of the pharmaceutical exposome in aquatic fauna, *Environmental Pollution* 239, 2018: 129-46.

Mompelat, B. and O. Thomas. Occurrence and fate of pharmaceutical products and by-products, from resource to drinking water, *Environment International* 35, 2009: 803-14.

Muhamedagic, Belma, et al. Dental office waste – public health and ecological risk, *Materia Socio Medica* 21(1), 2009: 35-8.

Nunes, Chalger, et al. Are there pharmaceutical compounds in sediments or in water? Determination of the distribution coefficient of benzodiazepine drugs in aquatic environment, *Environmental Pollution* 251, 2019: 522-9.

Orlando, Edward, et al. Endocrine-disrupting effects of cattle feedlot effluent on aquatic sentinel species, the fathead minnow, *Environmental Health Perspectives* 112, 2004:353-8.

Palma, Patricia, et al. Pharmaceuticals in a Mediterranean basin: the influence of temporal and hydrological patterns in environmental risk assessment, *Science of the Total Environment* 709, 2020: 136205.

Pivetta, Rhannanda, et al. Tracking the occurrence of psychotropic pharmaceuticals in Brazilian wastewater treatment plants and surface water, with assessment of environmental risks, *Science of the Total Environment* 727, 2020:138661.

Puckowski, Alan, et al. Bioaccumulation and analytics of pharmaceutical residues in the environment: a review, *Journal of Pharmaceutical and Biomedical Analysis* 127, 2016: 232-55.

Qiu, Wenhui, et al. Single and joint toxic effects of four antibiotics on some metabolic pathways of zebrafish (*Danio rerio*) larvae, *Science of the Total Environment* 716, 2020: 137062.

Rehman, Muhammad, et al. Global risk of pharmaceutical contamination from highly populated developing countries, *Chemosphere* 2013: 02.036.

Reports and Data. Medical waste management market to reach USD 17.89 billion by 2026, *Reports and Data*, November 26, 2019.

Reuters. US drinking water contamination with "forever chemicals" far worse than scientists thought, *The Guardian*, January 22, 2020.

Richmond, Erinn, et al. A diverse suite of pharmaceuticals contaminates stream and riparian food webs, *Nature Communications* 9, 2018: 4491.

Richtel, Matt and Andrew Jacobs. A mysterious infection, spanning the globe in a climate of secrecy, *New York Times*, April 6, 2019.

Ross, Daniel. Rainwater in parts of US contains high levels of PFAS chemical, says study, *The Guardian*, December 17, 2019.

Sabanoglu, Tugba. Total number of retail prescriptions filled annually in the U.S., 2013-2025, Statista, downloaded June 5, 2021.

Samuel, Sigal. The post-antibiotic era is here, *Vox*, November 14, 2019.

Schug, Thaddeus, et al. Minireview: Endocrine Distuptors: Past Lessons and Future Directions, *Molecular Endocrinology* 30(8): 2016.

Smith, Charlotte. Managing Pharmaceutical Waste, *Journal of the Pharmacy Society of Wisconsin*, Nov/Dec 2002.

Sneed, Annie. Forever Chemicals Are Widespread in U.S. Drinking Water, *Scientific American* January 22, 2021.

Subbaraman, Nidhi. These Five Brands of Dental Floss May Expose People to Harmful Chemicals, Study Finds, *Buzzfeed News*, January 9, 2019.

Usui, Noriko, et al. Assessment of the acute toxicity of 16 veterinary drugs and a disinfectant to aquatic and soil organisms, *Fundamental Toxicological Sciences* 6(9), 2019: 333-40.

Verlicchi, P and M Al Aukidy and E Zambello. Occurrence of pharmaceutical compounds in urban wastewater: removal, mass load and environmental risk after a secondary treatment – a review, *Science of the Total Environment* 429, 2012: 123-55.

Wang, Aolin, et al. Suspect Screening, Prioritization, and Confirmation of Environmental Chemicals in Maternal-Newborn Pairs from San Francisco, *Environmental Science and Technology* March 16, 2021.

Weisberg, Jessica. Death's Best Friend, *NY Review of Books*, April 2, 2018.

Williams, Hywel and Timothy Lenton. Microbial Gaia: a new model for the evolution of environmental regulation, Earth System Modeling Group, School of Environmental Sciences, September 19, 2007.

World Health Organization. Health Care Waste, WHO, February 8, 2018.

World Health Organization. Management of Solid Health-Care Waste at Primary Health-Care Centers, WHO, 2005.

Yoshida, Kate. Anti-anxiety drugs in wastewater impact fish behavior, *Ars Technica*, February 14, 2013.

Zanolli, Lauren. Bisphenol: what to know about the chemicals in water bottles and cans, *The Guardian*, May 24, 2019.

Zanolli, Lauren. Phthalates: why you need to know about the chemicals in cosmetics, *The Guardian*, May 23, 2019.

To learn more about Stephen's work see

www.stephenharrodbuhner.com
www.gaianstudies.org

Upcoming books from Raven Press

The Confluence of Gaia, Plant Medicine, and the Human Soul
by Julie McIntyre, with a foreword by Stephen Harrod Buhner.
Coming August, 2022.